Jesús Hazael García Gallegos
Mayola Giselle Galvan Mondragon
Carlos Manuel Dorantes Ayala

Integrated Water Purification System in Rural Communities

Jesús Hazael García Gallegos
Mayola Giselle Galvan Mondragon
Carlos Manuel Dorantes Ayala

Integrated Water Purification System in Rural Communities

Development of a solar prototype capable of generating clean water without harmful microorganisms and at low cost.

ScienciaScripts

Cover image: www.ingimage.com

This book is a translation from the original published under ISBN 978-620-0-02196-0.

Publisher:
Sciencia Scripts
is a trademark of
Dodo Books Indian Ocean Ltd. and OmniScriptum S.R.L publishing group

120 High Road, East Finchley, London, N2 9ED, United Kingdom
Str. Armeneasca 28/1, office 1, Chisinau MD-2012, Republic of Moldova, Europe
Managing Directors: Ieva Konstantinova, Victoria Ursu
info@omniscriptum.com

Printed at: see last page
ISBN: 978-620-8-62361-6

AUTHORS

JESÚS HAZAEL GARCÍA GALLEGOS MAYOLA GISELLE GALVÁN MONDRAGÓN CARLOS MANUEL DORANTES AYALA JUAN GABRIEL RODRÍGUEZ ORTIZ JOSÉ GASPAR BARRÓN OSORNIO JORGE ALBERTO CALLEJAS RUIZ

INDEX

INTRODUCTION

In response to the urgent need for access to safe drinking water in marginalised communities, a solar disinfection prototype for the treatment of stagnant and rainwater was designed and developed. This project focused on providing an economical and sustainable solution that could be implemented in rural and peri-urban areas, where resources are limited and access to advanced technologies is restricted. The prototype was developed with the aim of using solar energy, an abundant and free source of energy, to effectively disinfect water, providing a viable and accessible alternative to traditional water purification methods.

To ensure the effectiveness of the prototype, extensive verification tests were carried out, including Chemical Oxygen Demand (COD), suspended solids measurement, pH and electrical conductivity of the treated water. These tests were instrumental in assessing the system's ability to remove pollutants and improve water quality. The results obtained showed a significant reduction in the pollutant load, confirming that the prototype met the quality standards required for human consumption.

In addition, the ergonomics of the prototype were assessed, ensuring that its design was accessible and easy to operate by the target communities. The final design not only proved to be efficient in terms of performance, but was also found to be practical and adaptable to local conditions, enabling its implementation with low production and maintenance costs. These results highlight the viability of the prototype as an effective solution to improve access to drinking water in vulnerable areas, thus contributing to improving the quality of life of its inhabitants.

CHAPTER I

GENERAL INFORMATION ABOUT THE PROJECT

1.1 Title of the project

Integrated Water Purification System in Rural Communities.

1.2 Problematic

Access to clean water is a critical issue affecting numerous communities around the world, especially in rural areas and impoverished regions. The consequences of lack of access to clean water are diverse and profound, affecting the health, economy, education and environment of affected communities. Waterborne diseases: Lack clean water is directly linked to the spread of diseases such as cholera, dysentery, typhoid fever and diarrhoea. According to the World Health Organisation (WHO), diarrhoeal diseases caused by contaminated water kill around 485,000 people each year, many of them children under the age of five. Of those affected by malnutrition, chronic diarrhoea and other water-related diseases can lead to malnutrition, especially in children, due to the continued loss of essential nutrients and the body's inability to absorb food properly. While the economic impact does not have sufficient resources to generate a pipeline of clean water for consumption, loss of productivity and lack of access to clean water forces people, especially women and children, to spend several hours a day collecting water from distant sources. This time could be used for productive activities, such as paid work, education or family care. Given the factors of consuming unreliable water, medical costs, which are illnesses caused by contaminated water, generate significant medical costs for families, many of whom already live in poverty. These costs can further aggravate their economic situation.

1.3 Objective of the project

Design a prototype that meets the needs of solar disinfection of stagnant and/or rainwater, targeting rural communities with little access to water.

1.4 Justification

The development of a prototype capable of generating clean water free of harmful microorganisms at low cost is crucial to address the crisis of access to safe drinking water that affects millions people around the world. This project not only has the potential to improve the health and quality of life of vulnerable communities, but can also contribute to the economic and educational development of these regions. From which we would have a reduction of waterborne diseases such as cholera, dysentery and diarrhoea, which are major causes of morbidity and mortality in communities without access to safe drinking water. An effective prototype could drastically reduce the incidence of these diseases, improving public health and nutrition which reduces the prevalence of gastrointestinal diseases, the prototype would contribute to better nutrient absorption and thus reduce malnutrition, especially in children. Another advantage of this prototype is the economic impact at the time of purchase, it is low cost and saves on medical costs, which would free up economic resources that could be used for other basic needs or productive investments. While the environmental impact which generates sustainability designed to be efficient and low cost could utilise sustainable technologies, such as solar energy or rainwater harvesting systems, reducing dependence on overexploited water sources which would reduce pollution which by providing clean and safe water reduces the need to resort to polluted sources and pressure on the local ecosystem, helping to conserve the environment. By designing a low-cost prototype is essential for adoption in low-income communities, this will ensure that the system is affordable and can be implemented on a large scale and has an uncomplicated user interface for better understanding and use without the need for advanced skills.

1.5 Outreach

The scope of this project is to generate a prototype in which the communities that would benefit from it will be selected, based on their lack of access to safe drinking water and reliance on stagnant and rainwater sources; also by conducting water quality studies which would analyse the characteristics of the water in these areas, including common contaminants and levels of pathogenic microorganisms. Construction of a first working prototype using the selected materials and components, initial evaluation tests of the prototype in a controlled environment to verify its effectiveness in removing micro-organisms and contaminants, optimising the design refinement based on the results of the initial tests, optimising performance and reducing costs where possible in which the development of a conceptual design of the prototype, considering factors such as capacity, ease of use, durability, cost effectiveness and maintainability, will use materials and components from a selection of accessible and low-cost materials and components that can be obtained locally or easily distributed.

CHAPTER II
THEORETICAL FRAMEWORK

2.1 Prototype

Prototypes function as tangible representations of design ideas, allowing designers to turn their ideas into reality and test them in practice during the research and design phase. By creating prototypes, designers can test the functionality, usability and overall user experience of a design before investing significant resources in full-scale development. This iterative process of testing and refining prototypes allows designers to identify and eliminate any flaws or areas for improvement, resulting in a more refined and successful final design. Meanwhile, prototypes also play an important role in improving the user experience by providing a realistic representation of the application or website design. Designers can test the interaction flow, navigation and visual elements to ensure a smooth and intuitive user experience. By customising the user experience through prototyping, designers can create interfaces that are attractive, easy to use and meet user expectations (Engineering, 2024).

2.2 Radiation solar

Solar radiation is the energy emitted by the sun through electromagnetic waves, on which life on earth depends. In addition to determining atmospheric and climatic dynamics and trends, it also enables plant photosynthesis. If you want to know more about what types of radiation exist and how harmful they are for your health, especially for your skin in summer, read on.

The energy emitted by the sun in the form of electromagnetic radiation reaching the atmosphere. It is measured on a horizontal surface, by means of the radiation sensor or pyrometer, which is placed facing south and in a shadow-free location. The unit of measurement is watts per square metre (W/m^2).The solar radiation measured at each of the weather stations is given in units of power and is in watts per square metre (W/m^2). In the case of the data collected every 10 minutes it is the average power in 10 minutes and in the case of the daily radiation it represents the average power of the day.

If you want to convert the global solar radiation in units of power to units of energy, in the case of using the 10-minute data, multiply each of the power values in W/m^2 by 600 s (seconds in 10 minutes) and the result will be in Joules per square metre (J/m^2). In case the daily average global solar radiation value is used, multiply the power value in W/m^2 by 86,400 s (seconds in a day) and the result will be in Joules per square metre (J/m^2).

Light is the radiation that is visible to the human eye. Its wavelength is between 400 and 730 nm. Radiation whose wavelength is less than 400 nm is called ultraviolet radiation, and radiation whose wavelength is greater than 730 nm is called infrared radiation (Navarra, 2022).

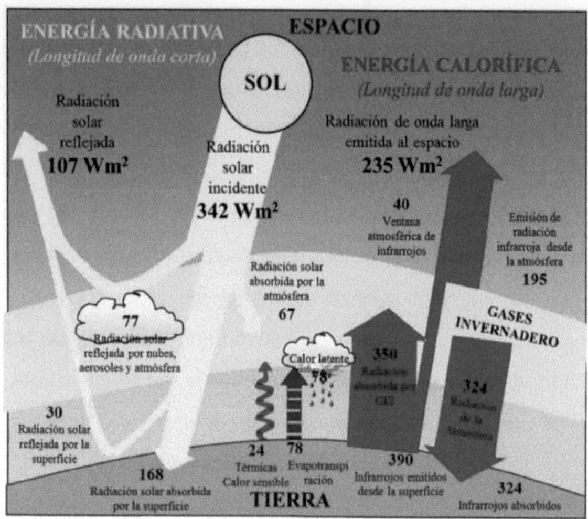

Figure 1. Types of energy into which solar radiation is transformed (Ferrer, n.d.).

2.3 Heat transfer by radiation

In radiation, energy is transmitted in the form of electromagnetic waves travelling at the speed of light. The electromagnetic radiation considered here is thermal radiation. The amount of energy leaving the surface in the form of radiant heat depends on the absolute temperature and the nature of the surface. An ideal emitter or black body emits from its surface a certain amount of radiant energy per unit time qr, determined by the equation. Radiation is the non-contact transfer of heat between objects. It is produced by the emission of energy through electromagnetic waves (Vega, 2004). Radiation is the transfer of heat via electromagnetic waves. It could be categorised as molecular transport, since the energy is produced by the change in the electronic configurations of the constituent atoms or molecules and transported by electromagnetic waves or photons. There is no direct contact between the two media and the intermediate or interface does not participate in the exchange functions - in most cases it is air, although there is also heat transfer through vacuum (UNAM, 2003).

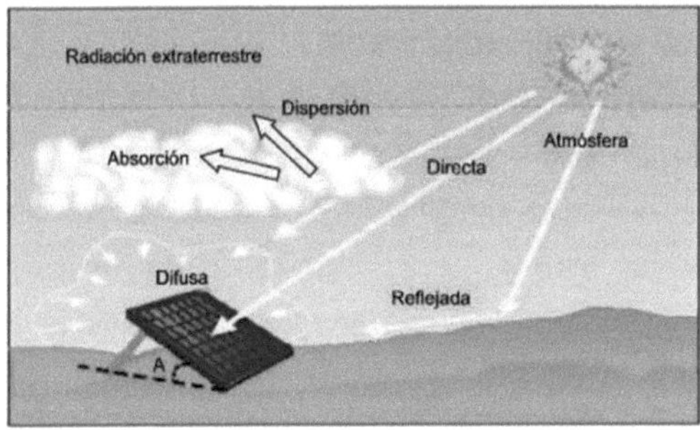

Figure 2. Energy transfer by solar radiation (UNEFA, n.d.).

2.4 Heat transfer by convection

Thermal convection involves the transfer of heat through the movement of a fluid, either liquid or gas. It can be classified into two types: natural convection and forced convection. Convective heat transfer is the process by which heat is transferred from one region to another through a moving fluid.

Natural Convection

Natural convection occurs when fluid movement is caused by density differences resulting from temperature variations in the fluid. For example, when air near a hot surface is heated, it expands and becomes less dense, causing it to rise and be replaced by cooler, denser air. This continuous cycle generates convection currents.

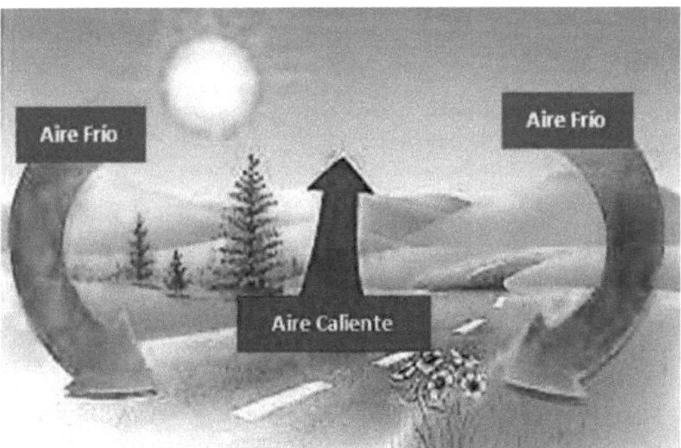

Figure 3. Natural convection process (Hsaini, Y., 2020.)

Forced Convection

In forced convection, fluid movement is induced by an external source, such as a fan, pump, or other mechanical device. This type of convection is common in heating and cooling systems, where a fan is used to move hot or cold air through a system. Similarly, using Newton's Law of Cooling, the heat transfer coefficient for forced convection is The forced convection heat transfer coefficient (h h) depends on several factors, including the fluid velocity, the fluid properties (density, viscosity, thermal conductivity, and heat capacity), and the geometry of the system. It is usually calculated using dimensionless numbers such as the Nusselt number (Nu), Reynolds number (R e), and Prandtl number (Pr) (Perez, 1984).

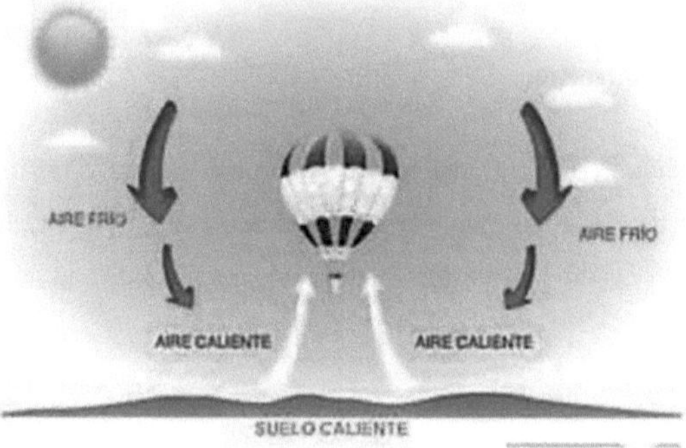

Figure 4. Energy by convection (Facilitator, P., n.d.).

2.5 Heat transfer by conduction

Heat transfer by conduction is a process in which heat is transferred through a solid or liquid material at rest under the influence of a temperature gradient. Conduction is the fundamental mechanism of heat transfer and can be described mathematically by Fourier's law (Energy and Combustion, 2009).

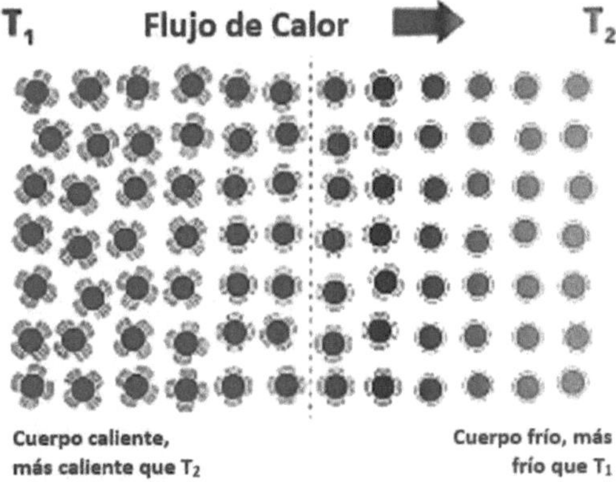

Figure 5. Heat transfer by convection (Shutterstock, 2024)

2.6 Fourier's Law

This law states that heat flows between two objects proportional to the temperature difference between them and can only flow in one direction: Heat can only transfer from a hotter object to a cooler object. In Theory of Thermal Analysis, Fourier, J. (1822, p. 2) states: "Many phenomena exist which are not caused by force mechanically, but only arise as a result of the presence and accumulation of heat. Heat flow is only possible between those regions that are at different temperatures, and its direction is always from the body of higher temperature to the body of lower temperature. The proportionality factor or constant is called the thermal conductivity of the material. There are various types of materials, from good conductors such as gold, silver or copper, to poor conductors called insulators such as glass or asbestos. The factor that determines how good a conductor a material is is thermal conductivity, which is defined as a physical property that measures the heat conducting capacity or ability to transfer the kinetic motion of its molecules to the molecules of adjacent bodies with which it in contact. The inverse of thermal conductivity is thermal resistance, which is the ability of a body to oppose the flow of heat (2018, Jimenez).

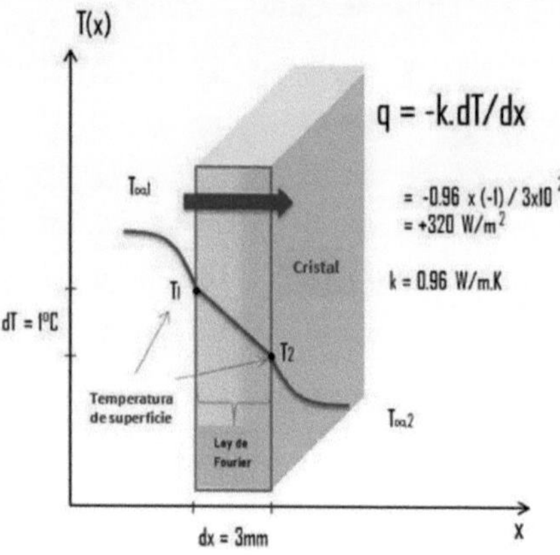

Figure 6. Heat transfer diagram, according to Fourier's Law (Therma Engineering, n.d.).

2.7 Temperature

Temperature is a scalar quantity defined as the amount of kinetic energy of particles of gaseous, liquid or solid mass. The higher the velocity of the particles, the higher the temperature and vice versa. Temperature measurement includes the concepts of cold (lower temperature) and heat (higher temperature), which can be felt instinctively. In addition, temperature also serves as a reference value for determining normal human body temperature, and is information used to assess health. Heat is also used in chemical, industrial and metallurgical processes (Leskow, 2024).

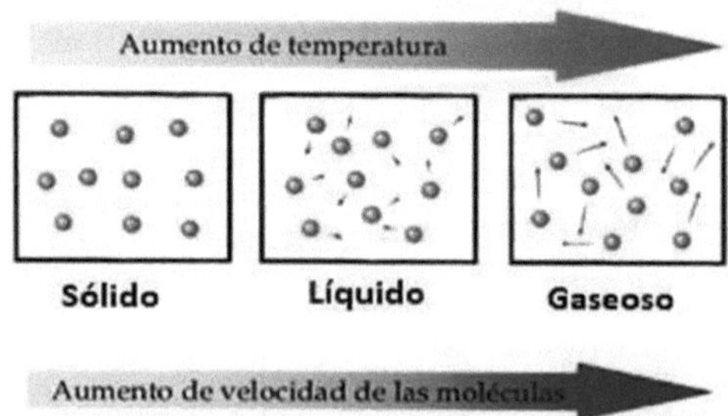

Figure 7. Diagram showing the movement of molecules according to the increase in heat supplied to a system (Wited, n.d.).

2.8 Scale of the temperature

The three most common temperature scales are: Celsius, Fahrenheit and Kelvin. A temperature scale can be created by identifying two easily reproducible temperatures. The boiling (change from liquid to vapour) and melting (change from solid to liquid) temperatures of water at one atmosphere of pressure. The Celsius scale. Also known as the "centigrade scale", it is the most commonly used along with the Fahrenheit scale. On this scale, the freezing point of water is 0 °C (zero degrees Celsius) and its boiling point is 100 °C (212 °F). The Fahrenheit scale. This is the measurement used in most English-speaking countries. On this scale, the freezing point of water occurs at 32 °F (thirty-two degrees Fahrenheit) and its boiling point at 212 °F (212 °C). The Kelvin scale. This is the measurement commonly used in science and establishes "absolute zero" as the zero point, which assumes that heat is given off by the object and is equivalent to -273.15 °C (degrees Celsius). Rankine scale. This is the commonly used measure in the United States for thermodynamic temperature measurement and is defined by measuring degrees Fahrenheit above absolute zero, so it has no negative or sub-zero values (Juan, 2018).

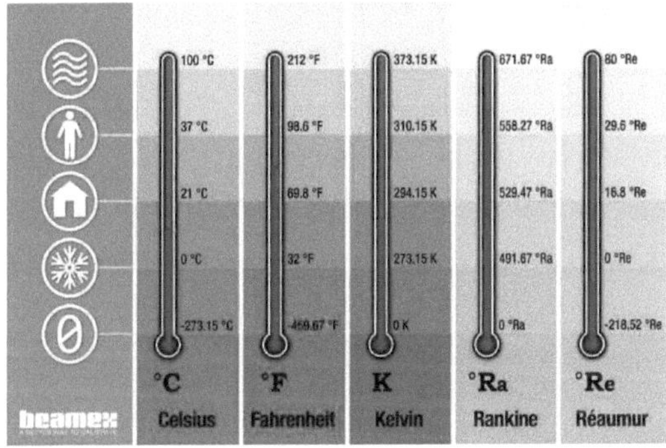

Figure 8. Different temperature scales and their equivalences (Beamex, n.d.).

2.9 Heat

Heat is the transfer of thermal energy due to temperature differences. This temperature difference is also known as a temperature gradient. Because heat is the movement of energy, it is measured in the same units as energy: joules (J), also called joules. It is also important to note that work and heat are closely related (Leskow, 2021).

Figure 9. Diagram of the relationship between the sun, energy and temperature (Docenteca, n.d.).

2.10 Irradiance solar

Solar irradiance is a key measure in studies related to solar energy, meteorology and climatology. It refers to the power per unit area received from the sun in the form of electromagnetic radiation at the top of the atmosphere or at the level of the earth's surface. Understanding solar irradiance is crucial for the design and optimisation of solar energy systems, as well for the analysis of weather and climate effects. Solar irradiance (E) is defined as the power density of solar radiation received by a surface per unit area and expressed in watts per square metre (W/m^2). This measure includes both direct radiation from the sun and diffuse radiation, which is radiation scattered by the atmosphere (Duffie & Beckman, 2013).

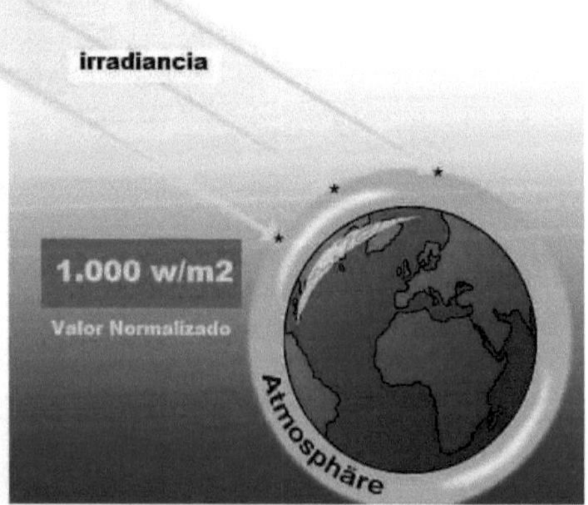

Figure 10. Diagram and normalised value solar irradiance (Área Tecnología, n.d.).

- Direct radiation: Radiation that arrives directly from the sun without scattered or reflected.

DNI

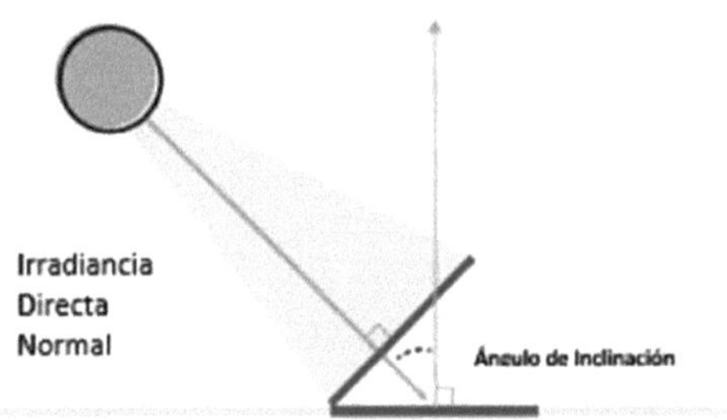

Figure 11. Diagram of Direct Irradiance (Solar Anywhere, 2021).

- Diffuse radiation: Radiation that has been scattered by molecules and particles in the atmosphere.

DHI

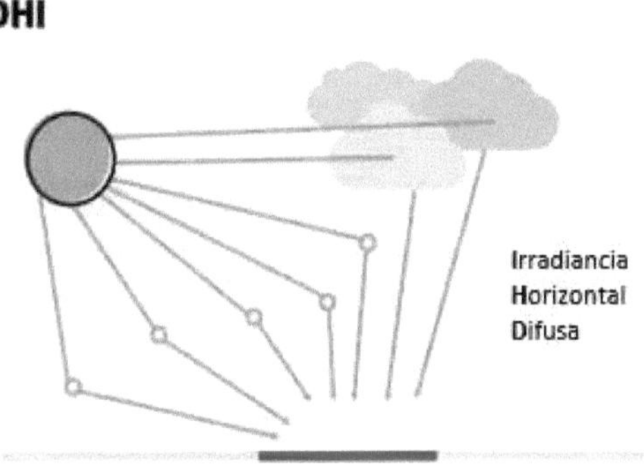

Figure 12. Horizontal diffuse irradiance diagram Solar Anywhere, 2021).

- Global Radiation: The sum of direct and diffuse radiation reaching the Earth's surface.

$CHI= DHI+ DNI * \cos(cx_{tilt})$

Figure 13. Equation for determining the Global Headroom Irradiance (GHI) (Solar Anywhere, 2021).

Factors Affecting Solar Irradiance:

- Latitude: Solar irradiance is highest in equatorial regions and decreases towards the poles.

Figure 14. Solar energy and latitude. (CK-12 Foundation, n.d.)

- Season: Due to the tilt of the earth's axis, irradiance varies with the seasons, being higher in summer and lower in winter in each hemisphere.

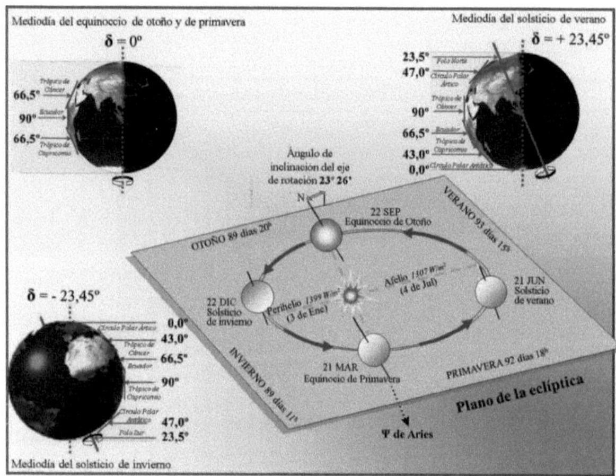

Figure 15. Solar energy and the seasons of the year (Ferrer, n.d.).

- Time of Day: Solar irradiance is highest at noon when the sun is at its highest point in the sky.

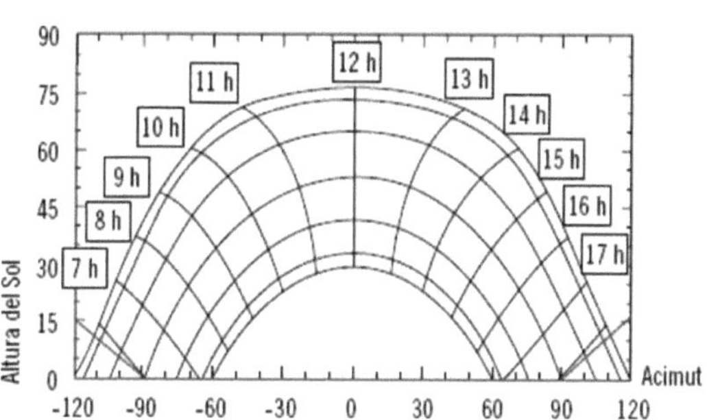

Figure 16. Graph showing the fluctuation of solar energy and the hours of the day (DBP, n.d.).

• Altitude: The higher the altitude, the thinner the atmosphere, which reduces the amount of scattered and absorbed radiation.

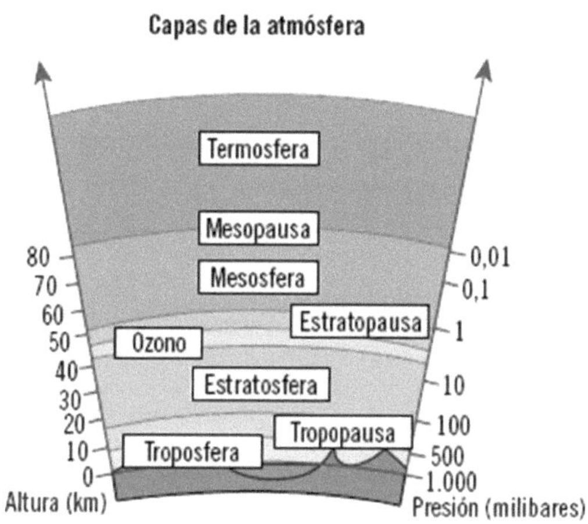

Figure 17. Layers of the atmosphere (DBP, n.d.)

• Atmospheric conditions: The presence of clouds, dust, and pollutants in the air can reduce the amount of solar irradiance reaching the earth's surface (Iqbal, 1983).

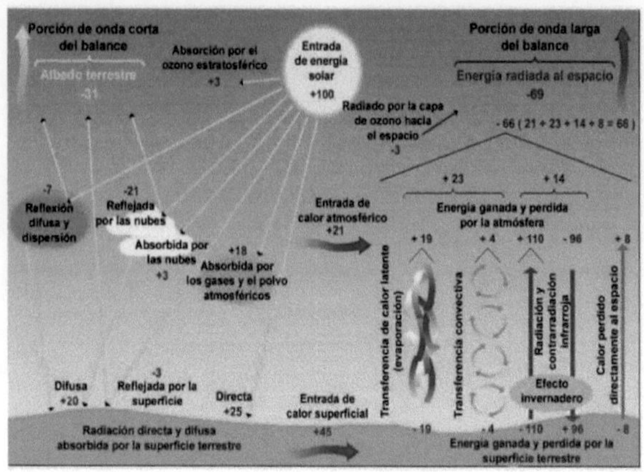

Figure 18. Energy balance of the atmosphere (Meteorología en Red, n.d.).

Solar Irradiance Measurement:

Accurate measurement of solar irradiance is essential for solar energy and climatology applications. Commonly used instruments include:

- Pyranometers: They measure the global solar radiation (direct+ diffuse) on a flat surface.

Figure 19. Photograph of a pyranometer (HACH, 2024).

- Pyrheliometers: Measure direct solar radiation.

Figure 20. Photograph of a Pyrheliometer (Elemetrics, 2024).

- Actinometers: Measure solar and other radiation.

Figure 21. Photograph of an Actinometer (Kipp and Zonen, 2024).

Solar Irradiance Applications

- Solar Photovoltaics: The efficiency and design of solar panels depend on the amount of solar irradiance available at a specific location (Masters, 2004).

Figure 22. Application of solar irradiance, for the generation of electrical energy by photovoltaic panels (Factor Energía, n.d.).

- Climatology and Meteorology: Solar irradiance affects climate and weather conditions by influencing temperature and the hydrological cycle (Stull, 1988).

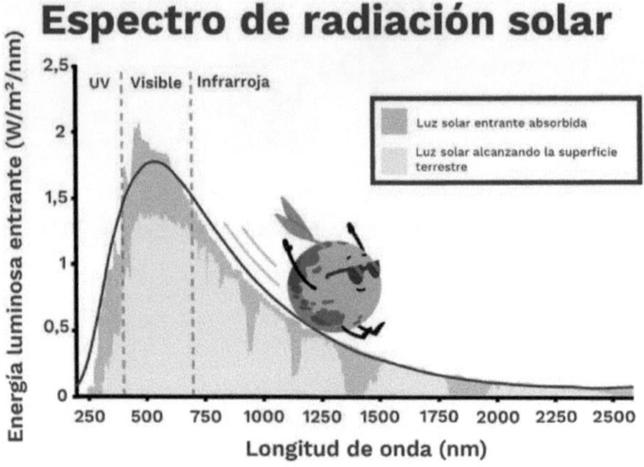

Figure 23. Solar radiation spectrum (Climate Science, n.d.).

- Agriculture: The amount of solar irradiance influences photosynthesis and thus plant growth and agricultural productivity (Monteith & Unsworth, 2013).

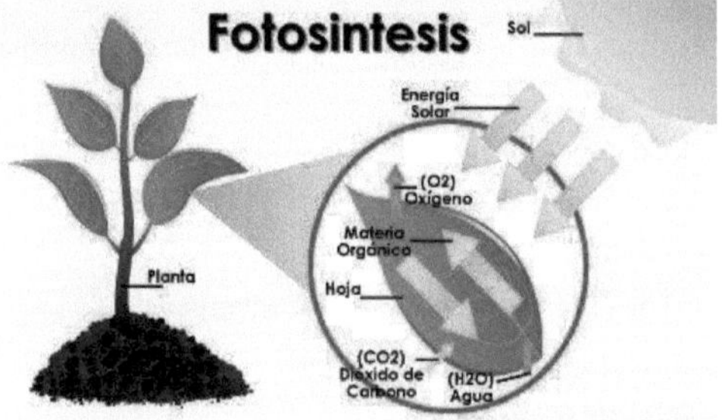

Figure 24. Importance of solar irradiance on photosynthesis and seedling production (Proain, n.d.).

2.11 Focal point of the evacuated tubes of a solar heater

The focal point in a solar heater is the place where the sun's rays converge after being reflected or refracted by a parabolic surface or lens. This principle is used in several types of solar heaters, including concentrating collectors, to maximise the efficiency of solar energy collection (Duffie & Beckman, 2013).

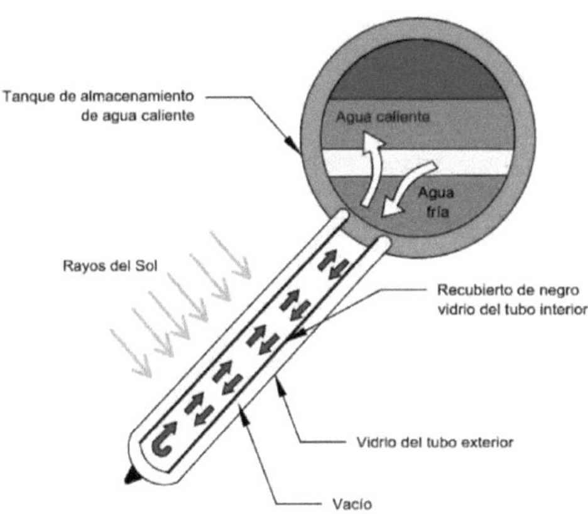

Figure 25. Operation of a evacuated tube solar water heater Renewable Alternative, 2016).

Parabolic collectors: Use parabolic mirrors to concentrate the sun's rays onto a receiver tube located at the focal point of the parabola. The fluid flowing through the tube absorbs the heat and transfers it to a storage system or directly to a point of use (Kalogirou, 2014).

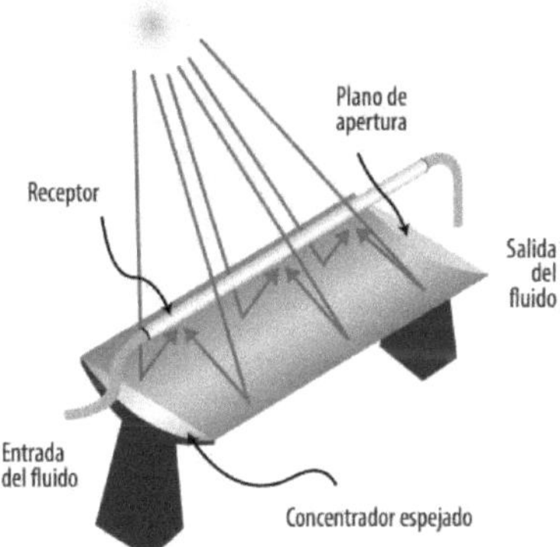

Figure 26. Operation of parabolic troughs (Ávila, J., 2017).

Fresnel collectors: These use an array of flat mirrors arranged at an angle to focus the sun's rays onto a fixed receiver. These collectors also rely on the focal point principle to increase the solar energy density at the receiver (Lovegrove & Stein, 2012). Applications of the focal point in solar heaters are shown below.

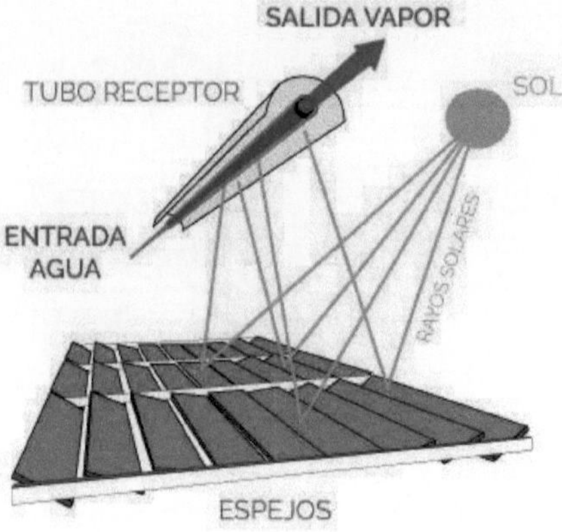

Figure 27. Operation of Fresnel collectors (RESSSPI, n.d.).

Electricity Generation: Concentrating solar power systems using focal points are essential in solar thermal power plants. These systems heat a thermal fluid that is then used to generate steam and drive turbines for electricity production (Zarza et al., 2004).

Figure 28. Central tower for power generation (Keeui, 2021).

Water Heating: In domestic and industrial applications, concentrating solar are used to heat water to high temperatures, reducing dependence on fossil fuels and lowering energy costs (Mekhilef et al., 2011).

Figure 29. Photograph and operation of a solar water heater. Representing hot water in red and cold water in blue (Ciencia UNAM, n.d.).

Industrial processes: Processes requiring high temperature heat, such as desalination, pasteurisation and sterilisation, can benefit from the use of solar heaters with focal points for a sustainable and efficient heat source (Horta, Mendes & Monteiro, 2018).

Figure 30. Design of a heat pump industrial heating (Ingelcia, n.d.).

2.12 Concentrator solar

A solar concentrator is an optical system that focuses incident solar radiation on a small area, increasing its power density. There are several types of solar concentrators, each with its own specific characteristics and applications (Kalogirou, 2014).

- Linear concentrators: These include parabolic trough and linear fresnel collectors. These systems concentrate sunlight along a focal line.

Figure 31. Linear concentrating solar power system (Eurostar Solar, n.d.).

- Point concentrators: These include parabolic dishes and heliostats in solar tower plants. These systems concentrate sunlight on a focal point.

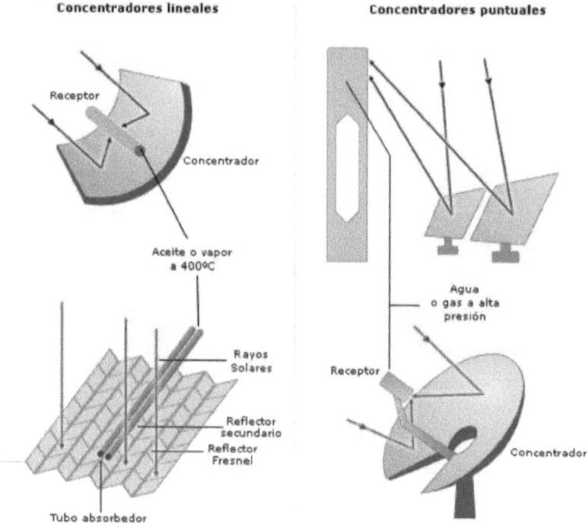

Figure 32. Schematic of point concentrators (The Morning Star G2, n.d.).

- Nonimaging concentrators: Such as Compound Parabolic Concentrator (CPC) type concentrators, which do not form images, but are efficient in capturing diffuse and direct solar radiation.

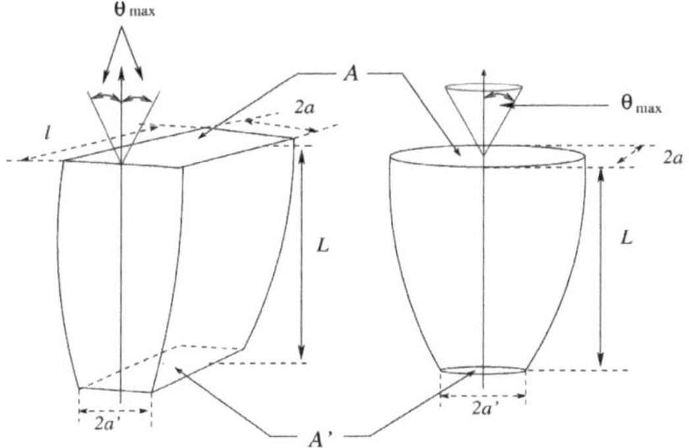

Figure 32. Perspective diagram of Nonimaging concentrators, a) Compound Parabolic Concentrator (CPC) in 2D, b) CPC in 3D (S., S. & Del Río, Jesus, 2009).

Types of Solar Concentrators Cylindrical-Parabolic Collectors:

They use parabolic mirrors to concentrate sunlight along a receiver tube. Commonly used in solar thermal power plants.

Capable of reaching high temperatures, suitable for generating steam and electricity (Duffie & Beckman, 2013).

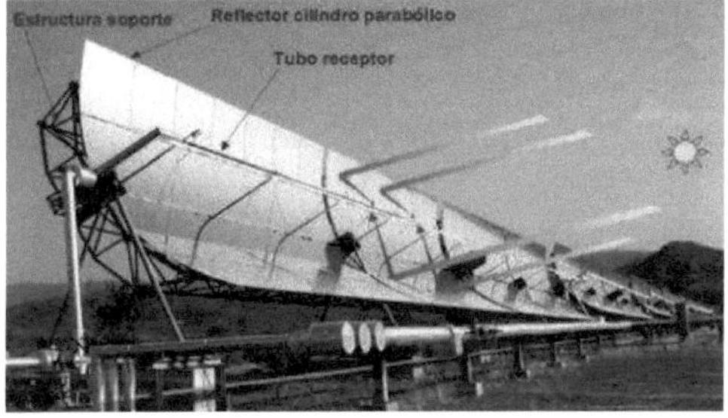

Figure 33. Parabolic trough concentrator technology (The Morning Star G2, 2012).

Parabolic Discs:

They use a parabolic reflector to focus sunlight onto a receiver located at the focal point. High concentration efficiency, suitable for applications requiring high temperatures (Wagner & Gilman, 2011).

Figure 34. Photograph of a parabolic dish solar concentrator (Solar Concentration, n.d.).

Heliostats and Solar Towers:

They use flat mirrors (heliostats) to direct sunlight to a receiver located on top of a tower. Used in tower solar power plants, capable of large-scale electricity generation (Kalogirou, 2014).

Figure 35. Photograph of a solar concentrator made up of heliostats (GMD Sol, n.d.).

Fresnel concentrators:

They use multiple flat mirrors to concentrate sunlight onto a linear receiver.

Simpler and less expensive design compared to parabolic troughs (Pihl & Boulay, 2012).

Figure 36. Fresnel solar concentrator (Hogarsense, n.d.)

Concentrators of Type CPC:

They capture both direct and diffuse radiation. They do not form images, but are very efficient at capturing light (Rabl, 1985).

Figure 37. Solar concentrator type CPC Fordecyt-IER UNAM, n.d.)

Principles of Operation

The principle of operation of solar concentrators is based on the laws of optics, specifically the reflection and refraction of light. The key components of a solar concentrator include:

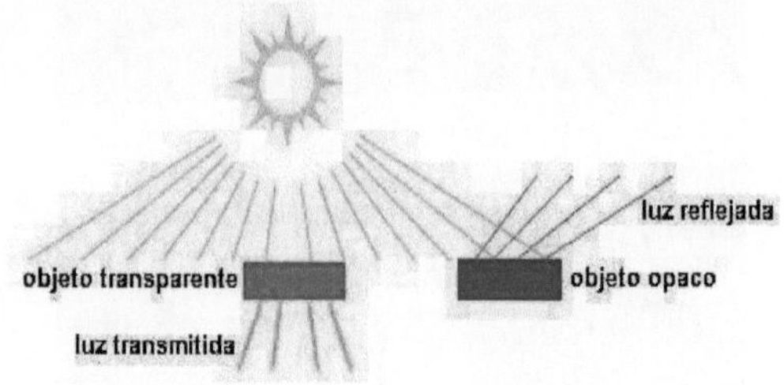

Figure 38. Reflection and refraction of light (Medium, 2019).

- Reflectors: Mirrored surfaces that reflect and concentrate sunlight onto the receiver.

Figure 39. Giant mirrors used as reflectors in winter in Norway (ArchDaily, 2013).

- Receivers: Surfaces or devices that absorb concentrated solar energy and convert it into heat or electricity.

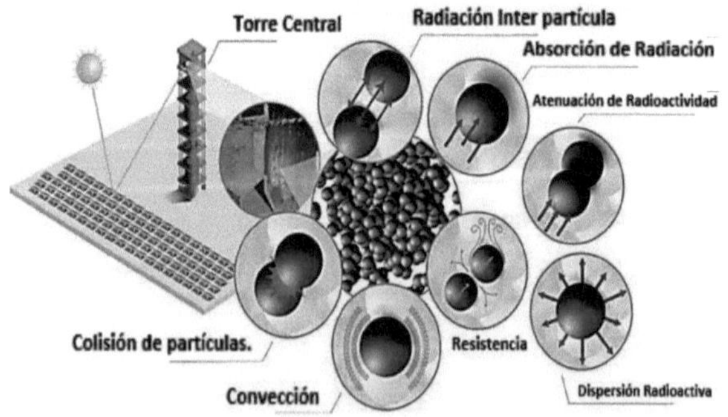

Figure 40. Conceptualisation of the new generation of concentrating solar power particle receivers (Ecoinventos, 2018).

- Solar Tracking: Systems that adjust the orientation of the concentrator to follow the path of the sun, maximising energy capture (Masters, 2004).

Figure 41. Solar tracking system (Solar Energy, n.d.).

2.13 Concentrator Fresnel

Fresnel concentrators are an advanced technology in the field of solar energy that allows the concentration of sunlight by means of multiple flat mirrors or prisms. Its simplified design and their ability to reduce manufacturing and maintenance costs make them attractive for a variety of solar thermal and photovoltaic applications. These concentrators are named after the French physicist Augustin-Jean Fresnel, known for his contributions to optics. A Fresnel concentrator is an optical system that uses multiple flat mirrors (or lenses) arranged concentrate sunlight onto a linear or point receiver. The main advantage of Fresnel concentrators is their modular design and lower cost compared to traditional parabolic systems (Kalogirou, 2014).

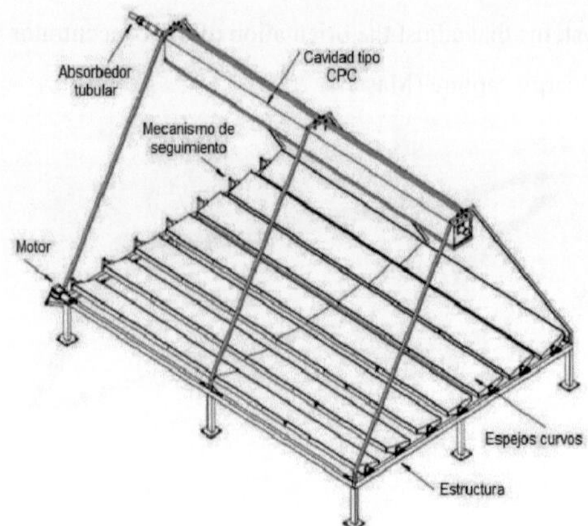

Figure 42. Linear Fresnel concentrator with curved mirrors and CPC cavity (Lara, F. et al, 2012).

Linear Fresnel Reflectors:

They use a series of flat mirrors to concentrate sunlight onto a fixed receiver tube. They can track the movement of the sun along one axis (uniaxial tracking) or on two axes (biaxial tracking).

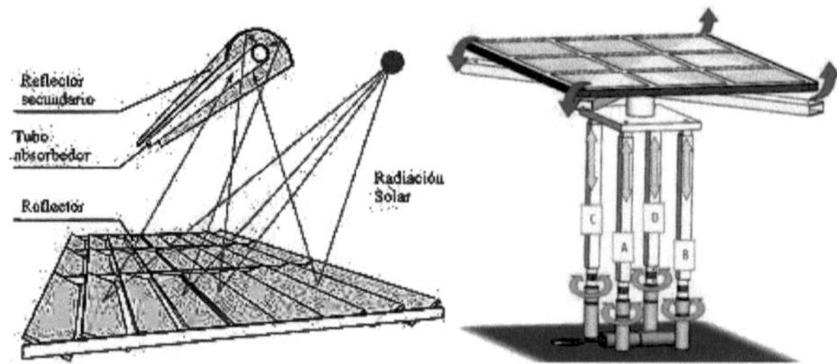

Figure 47. Linear Fresnel monoaxial and biaxial Fresnel reflector in order of appearance (Tecpa, n.d.).

Fresnel lenses:

They use a series of prisms or lens segments arranged in a circular pattern to concentrate sunlight. Commonly used in concentrated photovoltaic (CPV) applications.

Figure 48. Photograph the configuration of a Fresnel lens (Made in China, n.d.).

Principles of Operation

The principle of operation of Fresnel concentrators is based on the reflection and refraction of sunlight:

Flat Mirrors:

The flat mirrors are arranged in rows and each is set at a specific angle to direct sunlight towards a central receiver. The configuration of the mirrors allows the system is more compact and easier to manufacture than the parabolic concentrators (Pihl & Boulay, 2012).

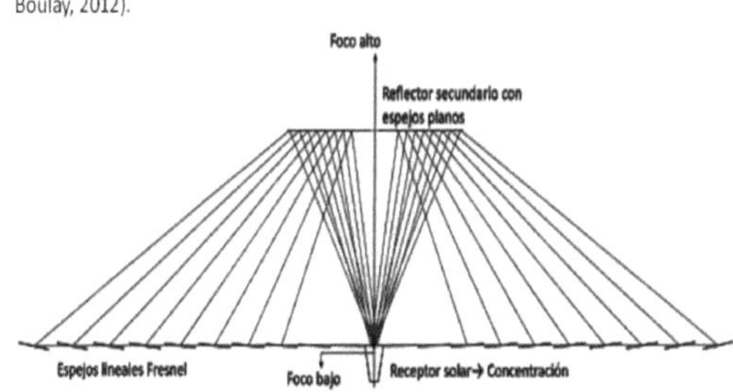

Figure 49. Diagram of configuration of plane reflection mirrors (Madrimasd, 2021).

Fresnel lenses:

Fresnel lenses are composed of a series of lens segments that are thinner than a conventional lens, reducing volume and weight. These lenses concentrate the sunlight into a small focal point, increasing the intensity of the radiation on the solar cells (Rabl, 1985).

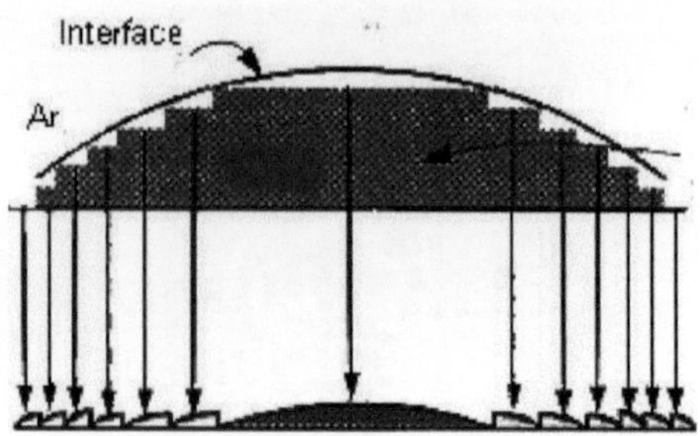

Figure 50. Diagram of the structure of a Fresnel lens (100cía en casa, 2012).

Advantages:

- Reduced Cost: Lower manufacturing cost due to the use of flat, easily manufactured materials.
- Modular Design: Allows for easy scalability and maintenance of the system.
- Design Flexibility: Can be integrated into various configurations and solar applications.

Disadvantages:

Lower Optical Efficiency: May have slightly lower optical efficiency compared to parabolic systems due to reflection and scattering losses.Solar Tracking Required: Requires an accurate solar tracking system to maximise energy capture.

Fresnel Concentrator Applications

Solar Thermal Power Plants:Used to generate steam to power turbines for electricity production. Suitable for large-scale applications due to their ability to concentrate large amounts of solar energy (Kalogirou, 2014).

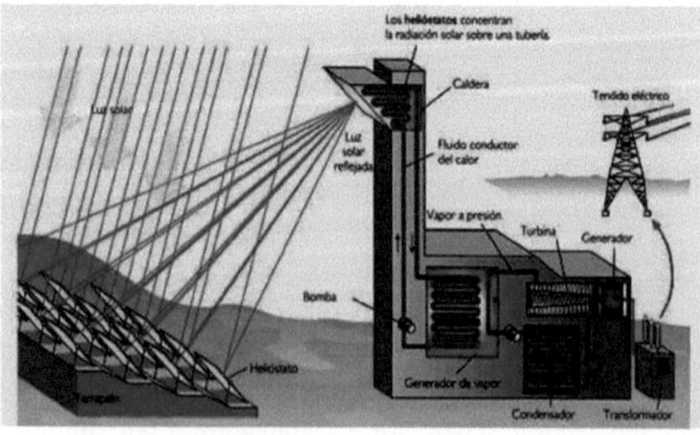

Figure 51. Solar thermal power plant (Canaltic, n.d.).

Concentrated Photovoltaic Systems (CPV):

Use of Fresnel lenses to focus sunlight on high-efficiency solar cells. It improves conversion efficiency and reduces the cost per watt of energy produced (Leutz & Suzuki, 2001).

Figure 52. Photographs of the application of concentrated photovoltaic (CPV) systems Eco Solar Esp., n.d.).

Fluid Heating:

Industrial applications for heating water or oil in processes requiring high temperatures. Benefits in terms of efficiency and reduced operating costs.
Challenges and Opportunities

Challenges:

- Solar Tracking Accuracy: The efficiency of the system is highly dependent on the accuracy of the solar tracking.
- Atmospheric Conditions: Variability in atmospheric conditions can affect concentration efficiency.

Opportunities:

- Technological Innovations: Development of reflective materials and more efficient lenses.
- Hybrid System Integration: Combination with other renewable energy technologies to improve reliability and energy production.

2.14 Characteristics of stagnant water and water from rainfall

Standing waters are bodies of water that at rest without significant flow. Common examples include ponds, ponds, puddles, and reservoirs where water does not circulate. These waters can arise from natural sources, such as rainfall accumulation, or from artificial sources, such as water collection in man-made structures (Smith, 2020.).

Figure 53. Photographs of standing water bodies (Tunes, S., 2020).

Characteristics:

- Lack of movement: The main characteristic of stagnant water is the absence of movement or circulation, which causes water to remain in one place for a prolonged period of time.

Figure 54. National Canal in Mexico City, showing water without movement (León, A., 2020).

- Proliferation of Microorganisms: Due to the lack of movement, these waters are prone to proliferation of bacteria, algae and other microorganisms that can affect their quality.

Figure 55. Standing water in Navojoa. (Castellón, J., 2022).

• Low Oxygen Level: Water stagnation often leads to a low level of dissolved oxygen, which can affect aquatic life and promote anaerobic conditions that lead to odours.

Figure 56. Lake where eutrophication problems are beginning to be perceived due to low levels of dissolved oxygen (Rodríguez, H., 2022).

• Contamination Risk: Standing water can accumulate pollutants, such as debris, chemicals and organic matter, which can make it unhealthy and dangerous for human and animal consumption.

Figure 57. River in Nigeria showing stagnation and pollution (Kashi, E., 2024).

Concept and Characteristics of Rainwater

Concept:

Rainwater, also known as stormwater, is the water that falls to earth during precipitation. This water can be collected and used for different purposes, such as irrigation, aquifer recharge, or, with proper treatment, human consumption (Garcia, 2017).

Characteristics:

- Seasonal Availability: Stormwater is available depending on weather patterns, making it a seasonal source of water.

Figure 58. Illustrative image of the availability water catchment in rainy season (Oliver, C., n.d.).

- Variable Quality: The quality of stormwater can vary significantly depending on the atmosphere, the surfaces on which it falls and the pollutants carried.

Figure 59. Acid rain formation by atmospheric pollution (Roucau, M., n.d.).

- Collection Potential: Stormwater can be collected from roofs and other hard surfaces, using catchment systems that allow for storage and subsequent use.

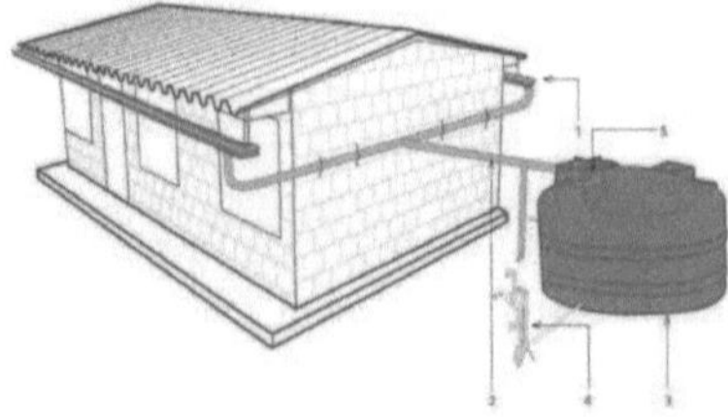

Figure 60. Rainwater collection implemented in (JAPAC, 2018).

- Lower Initial Pollution: Although it can carry pollutants from air and surfaces, stormwater generally has fewer pollutants than other water sources, such as surface water or groundwater.

Figure 61. Operation of rainwater collection channels, showing the crystallinity of the water collected (Tricanal, n.d.).

2.15 NOM-003-SEMARNAT- 1997

The Mexican Official Standard NOM-003-SEMARNAT-1997 aims to establish the maximum permissible limits of pollutants in treated wastewater that is reused in public services. This standard is essential to ensure that the reuse of treated wastewater does not represent a risk to public health and the environment.

This standard applies to:

- Public and private entities that treat and reuse wastewater in public services.
- Specific uses such as irrigation of green areas, golf courses, and in agriculture.
- Street and vehicle washing.
- Aquifer recharge, among others.

Parameters and Maximum Permissible Limits

The standard sets maximum permissible limits for various pollutants in treated wastewater. These limits are designed to protect public health and the environment. The main parameters include:

Microbiological:

- Faecal Coliforms: Less than 1,000 NMP/100 ml.
- Helminths: Less than 1 egg/L.

Physicochemicals:

- Total Suspended Solids (TSS): Maximum 30 mg/L.
- Biochemical Oxygen Demand (BOD5): Maximum 20 mg/L.

Heavy Metals and Other Compounds:

- Arsenic (As): Maximum 0.1 mg/L.
- Cadmium (Cd): Maximum 0.01 mg/L.
- Total Chromium (Cr): Maximum 0.1 mg/L.
- Lead (Pb): Maximum of 0.2 mg/L.
- Mercury (Hg): Maximum 0.001 mg/L.

Sampling and Analysis Methodology

NOM-003-SEMARNAT-1997 specifies the sampling and analytical methods to be used to measure the levels of pollutants in treated wastewater. These methods are consistent with national and international standards to ensure accuracy and reliability of results. Sampling: Sampling should be done representatively and at specific points in the wastewater treatment system. The use of standardised procedures is recommended to avoid contamination and deterioration of samples. Analysis: Analyses must be performed by accredited laboratories using validated methods. Results must be reported to the competent authorities according to established procedures (Secretaría de Medio Ambiente y Recursos Naturales, 2021).

2.16 NOM-001-SEMARNAT- 2021

The Official Mexican Standard NOM-001-SEMARNAT-2021 establishes the maximum permissible limits for pollutants in wastewater discharges into receiving bodies of national waters. Its objective is to protect the quality of water resources and preserve the environment and public health.

These are some of the applications of the standard:

- Stationary sources such as industrial, commercial, service or municipal sources.
- Any other source discharging waste water into receiving bodies of national waters.

Parameters and Maximum Permissible Limits

The standard defines maximum permissible limits for a variety of pollutants. These limits apply to wastewater discharges to rivers, lakes, lagoons, estuaries and marine waters, among other receiving bodies.

Physico-chemical parameters:

- Biochemical Oxygen Demand (BOD5): Maximum 75 mg/L.
- Total Suspended Solids (TSS): Maximum 75 mg/L.
- Chemical Oxygen Demand (COD): Maximum of 150 mg/L.
- Fats and Oils: Maximum 15 mg/L.
- pH: Between 5 and 10 units.

Heavy Metals and Other Pollutants:

- Arsenic (As): Maximum 0.2 mg/L.
- Cadmium (Cd): Maximum 0.01 mg/L.
- Copper (Cu): Maximum 4 mg/L.
- Total Chromium (Cr): Maximum 1 mg/L.
- Lead (Pb): Maximum of 0.2 mg/L.
- Mercury (Hg): Maximum 0.01 mg/L.

Other Relevant Parameters:

- Total Nitrogen: Maximum 15 mg/L.
- Total Phosphorus: Maximum 5 mg/L.
- Faecal coliforms: Maximum of 1,000 NMP/100 ml.

Sampling and Analysis Methodology

The standard details the procedures for sampling and analysis of wastewater to ensure the accuracy and reliability of the results:

Sampling: Sampling should be carried out representatively and at specific points in the discharge system. It requires the implementation of standardised procedures to avoid contamination and to ensure the integrity of the samples (Secretaría de Medio Ambiente y Recursos Naturales, 1997).

Analysis: Analyses must be performed by accredited laboratories using validated methods. Results must be reported to the Ministry of Environment and Natural Resources.

Natural Resources (SEMARNAT) and other relevant authorities (Secretaría de Medio Ambiente y Recursos Naturales, 2021).

2.17 NOM-127-SSA1- 2021

The Official Mexican Standard NOM-127-SSA1-2021 establishes the permissible quality limits and drinking water treatment for human use and consumption. This standard is crucial to ensure that the water supplied to the population is safe and suitable for human consumption, protecting public health.

This standard applies to:

- All water supply systems for human use and consumption, both public and private.
- Operators and providers of drinking water services, including municipalities, operators and concessionaires.

Parameters and Maximum Permissible Limits

NOM-127-SSA1-2021 specifies maximum permissible limits for a variety of microbiological, physical, chemical and radioactive contaminants. These parameters are essential to ensure the quality of drinking water.

Microbiological parameters:

- Total coliforms: 0 in 100 ml.
- Escherichia coli (E. coli): 0 in 100 ml.

Physico-chemical parameters:

- Turbidity: Not more than 5 NTU (Nephelometric Turbidity Units).
- pH: Between 6.5 and 8.5.
- True Colour: No more than 15 units of colour.

Heavy Metals and Other Chemical Contaminants:

- Arsenic (As): Maximum 0.025 mg/L.
- Cadmium (Cd): Maximum 0.003 mg/L.
- Chromium (Cr): Maximum 0.05 mg/L.
- Lead (Pb): Maximum 0.01 mg/L.
- Mercury (Hg): Maximum 0.001 mg/L.
- Nitrates (NO3): Maximum 10 mg/L.
- Fluorides (F): Maximum 1.5 mg/L.

Organic pollutants:

- Benzene: Maximum of 0.01 mg/L.
- Carbon tetrachloride: Maximum 0.002 mg/L.
- Chloroform: Maximum 0.07 mg/l.

Radioactive Parameters:

- Total Alpha: Maximum 0.1 Bq/L (Becquerel per litre).
- Total Beta: Maximum 1.0 Bq/L.

Sampling and Analysis Methodology

The standard details the procedures for sampling and analysis of drinking water to ensure the accuracy and reliability of the results:

Sampling: Sampling should be done representatively and at specific points in the supply system. Standardised procedures should be followed to avoid contamination and ensure sample integrity. Analysis: Analyses must be performed by accredited laboratories using validated methods. Results must be reported to the Ministry of Health and other relevant authorities (Ministry of Health, 2021).

2.18 NOM-230-SSA1- 2002

The Official Mexican Standard NOM-230-SSA1-2002 establishes the sanitary requirements that water supply systems for human use and consumption must comply with. Its main objective is to protect the health of the population by ensuring that the water supplied is safe and free of contaminants.

This standard applies to:

- All public and private drinking water supply systems.
- Operators and providers of drinking water services, including municipalities, operators and concessionaires.

Health requirements

NOM-230-SSA1-2002 specifies a number of requirements that water supply systems must meet to ensure water quality and safety. These requirements include:

Water Quality Control and Monitoring:

- Carry out microbiological, physical and chemical analyses of water at different points in the distribution system.
- Regularly monitor water quality to detect any deviation from established parameters.

Infrastructure and Operation:

- Catchment: Ensure that water sources (wells, springs, rivers) are protected from contamination.
- Treatment: Apply appropriate potabilisation processes, including coagulation, flocculation, sedimentation, filtration and disinfection.

- Storage: Maintain tanks and reservoirs in hygienic conditions and carry out regular cleaning.
- Distribution: Ensure that distribution networks are in good condition and free of leaks or cross-contamination.

Disinfection:

- Maintain a residual level of free chlorine in the distributed water between 0.2 and 1.5 mg/L to ensure continuous disinfection.
- Carry out regular checks of the chlorine level at different points in the distribution network.

Maintenance and Sanitation:

- Carry out preventive and corrective maintenance programmes on all supply system facilities.
- Implement basic sanitation measures in areas surrounding sources and treatment plants.

Training and Personnel:

- Train personnel responsible for the management and operation of water supply systems in good hygiene and water management practices.
- Having qualified personnel for the operation and maintenance of treatment and distribution systems.

Sampling and Analysis Methodology

The standard details the procedures for sampling and analysis of drinking water to ensure the accuracy and reliability of the results:

Sampling: Sampling should be carried out representatively at different points in the supply system, including the source, storage and the distribution network.

Analysis: Analyses must be carried out by accredited laboratories using standardised methods. The parameters to be analysed include total coliforms, residual chlorine, turbidity, pH, among others (Secretaria de la salud, 2002).

CHAPTER III
PROJECT DEVELOPMENT AND PROPOSED SOLUTIONS

3.1 Procedure for the elaboration of the disinfection prototype solar

1. A 20 L flexible plastic canister was painted with two coats of paint, the first with heat-resistant, quick-drying matt black paint (Truper R), the second coat was applied after drying, this time with heat-resistant, quick-drying glossy black spray paint. It was left to dry for 10 min.

Figure 62. Painting of carafe.

2. The cylinder was drilled with two circular holes with the aid of a drill and a 1/8 mm hole saw, which was measured at an inclination of 23°, the gaskets of the evacuated pipes were fitted, which prevented pressure leakage.

Figure 63. Drilling of a carafe

3. A PVC pipe of 8 in diameter and 2 m long was cut in half and a cut of 4 in was determined with a polishing machine with a 155 m disc (Pretul R). Then

After that, the two exterior of the tubes were painted with two coats of black paint, one matt and the other glossy, respecting the drying times of 3 min. These are called thermal concentrators.

Figure 64. Cutting and painting of tubes used as solar concentrators.

4. After the drying time of the tubes, the inside of the tubes were lined with aluminium (ready thermal concentrators).

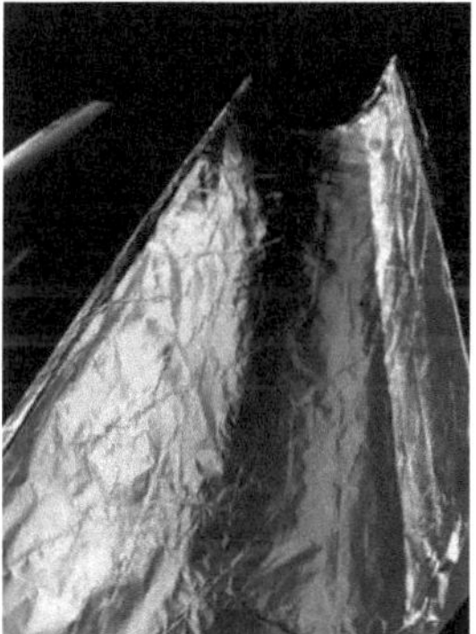

Figure 65. Foil lining of the concentrators.

5. After all the material preparation process, the assembly process of the pieces began, in which the carafe was placed on a 50 cm high base where the evacuated tubes were connected at 23° of inclination where they are mounted on the thermal concentrators in the same way. with a degree of inclination of 23° suspended from each other, resulting in an inclination of 23°.

Figure 66. Prototype assembly

6. After the assembly, water was added to a total of 30 L and connected to the distillate part of the tank where the distillation process was allowed to proceed and reached a temperature of > 100°.

Figure 67. Prototype start-up

7. Once the process or sunlight time was over, which is from 10 am to 4 pm, the water was removed and stored.

3.2 Determination of oxygen chemistry in natural, treated wastewater and treated wastewater (COD-TS).

In the framework of the development of the solar disinfection prototype for stagnant and storm water, a Chemical Oxygen Demand (COD) test was carried out, which was developed using the methodology of NMX-AA-030/2-SFCI-2011 to evaluate its efficiency in improving water quality. The COD test measured the amount of oxygen needed to oxidise the organic matter present in the water, providing an indication of the level of organic contamination. This test was carried out to verify the effectiveness of the prototype in removing contaminants and to ensure that the treated water meets health and safety standards. The results obtained allow the design of the system to be adjusted and optimised, ensuring that it is a viable and effective solution for communities with limited access to economic and technological resources.

- The following methodology was performed based on NMX-AA-030/2-SCFI-2011 in which a potassium dichromate solution, certified reference solution (where applicable), c (K_2Cr2O_7) = 0,10 mol/L (range up to 1 000 mg/L COD- TS) was prepared. Dissolve (29,418 ± 0,005) g potassium dichromate (dried at 105 °C for 2 h ± 10 min) in approximately 600 mL water in a beaker. To this 160 mL of concentrated sulphuric acid (see 6.4.1) was carefully added and stirred. It was then allowed to cool and diluted to 1 000 mL in a volumetric flask. The solution is stable for at least six months.

Figure 68. Potassium dichromate , used for preparation of COD vials.

- The following methodology was performed on the basis of NMX-AA-030/2-SCFI-2011 from which a solution of silver sulphate in sulphuric acid, c ($Ag(_2)SO_4$) = 0,038 5 mol/L was prepared. Dissolve (24,0 ± 0,1) g of silver sulphate in 2 L of concentrated sulphuric acid (see 6.4.1). Where a satisfactory solution was obtained, the initial mixture was stirred. It was then left to stand overnight and then stirred again in order to dissolve all the sulphate. It was stored in a dark glass bottle protected from direct sunlight. The solution is stable for twelve months.

Figure 69. Potassium dichromate solution, COD vials, and potassium acid phthalate standard solutions for calibration curve preparation.

• The following methodology was performed based on NMX-AA-030/2-SCFI-2011, in which a reference stock solution of phthalate mass concentration was prepared.

Potassium hydrogen phthalate (KHP) [C_6H_4 (COOH) (COOK)], and (COD-TS) of 10 000 mg/L. Dissolve (4,251± 0,002) g potassium hydrogen phthalate, previously dried at (105 ± 5) °C for 2 h ± 10 min in approximately 350 mL of water (see 6.1). It was diluted with water to 500 mL in a volumetric flask. The dilution was then stored under refrigeration at 2 °C to 8 °C and new dilutions were prepared every month. This dilution is commercially available. 6.8.2 Reference solutions for instrumental calibration, with mass and (TSS-CCD) concentration values of 200 mg/L, 400 mg/L, 600 mg/L, 800 mg/L and 1 000 mg/L. 20 mL, 40 mL, 60 mL, 80 mL and 100 mL of the 10 000 mg/L mass concentration reference stock solution (see 6.8.1) were diluted separately with 4 mL of dilute sulphuric acid (see 6.4.2) to 1 000 mL with water. Store these dilutions at 2 °C to 8 °C. °C and prepare new dilutions every month (when applicable).

Figure 70. Potassium hydrogen phthalate standard solutions for calibration curve preparation.

• The following methodology was performed based on NMX-AA-030/2-SCFI-2011 in which a mixture of (0,50 ± 0,01) mL potassium dichromate (see 6.3) was prepared in individual digestion tubes (see 7.1.2). Carefully added (0,20 ± 0,01) mL of mercury (II) sulphate solution (see 6.5), followed by (2,50 ± 0,01) mL silver sulphate (see 6.6). The tubes were carefully shaken and then capped. They were allowed to stand overnight to cool. It was then shaken again before use. This prepared reagent is stable for one year stored in a dark place at room temperature.

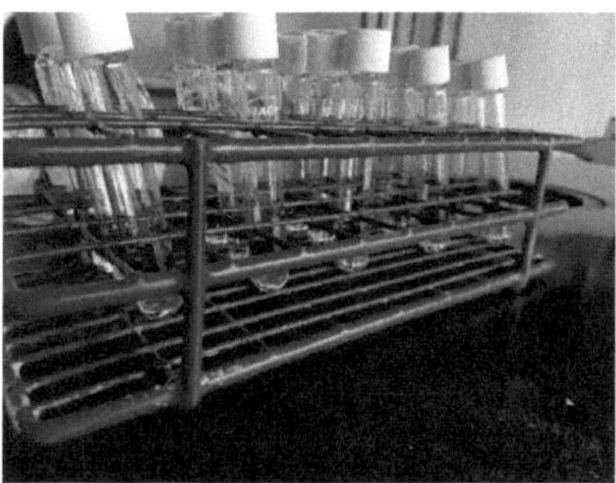

Figure 71. Vials for COD determination prepared in the laboratory.

• The following methodology was carried out based on the NMX-AA-030/2-SCFI-2011 in which the tubes were selected from 3 tubes for each solution giving a total of 15 tubes in which 6.5 ml of the respective solution from 200 mg/L to 1000 mg/L was added to each tube. The tubes were then shaken for a final time and placed in the thermo reactor previously preheated to 105 °C. After placing the dead cells together with a blank, they were left in the thermo reactor for 2 hours.

Figure 72. Addition of potassium hydrogen phthalate at different concentrations in vials for COD determination.

• The following methodology was performed based on the NMX-AA-030/2-SCFI-2011 where after 2 hours the samples were taken from the thermo reactor in which a more intense colour change was observed to the initial colour before putting it into the thermo reactor in which the samples were waited to cool down and ABS was measured in a colorimeter and the results were noted.

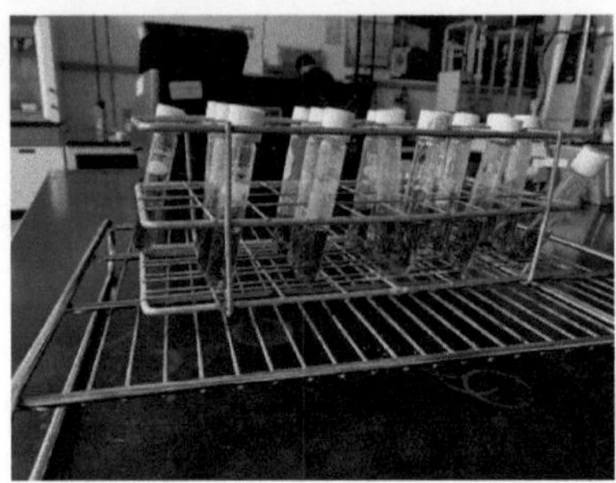

Figure 73. COD used in calibration curve, after thermal digestion.

• First we calibrated our colorimeter with our blank after the reading of 0.000 ABS is ready to measure the readings of the samples then we rinsed between each of the readings so we rinsed the container 15 times and between intervals of 3 samples we recalibrate our colorimeter. The methodology shown is based on NMX-AA-030/2-SCFI-2011.

Figure 74. HACH UV equipment, for reading ABS.

• After taking all the readings, a calibration curve was made with the average of every 3 readings of the different solutions, which resulted in 99.98, using the data from the graphical table of COD calibration. Methodology based on NMX-AA-030/2-SCFI-2011.

1000 mg/L	800 mg/L	600 mg/L	400 mg/L	200 mg/L
0.659	0.556	0.579	0.383	0.340
0.662	0.612	0.420	0.392	NA
0.645	0.534	0.446	NA	0.256

Figure 75. ABS readings, calibration curve.

1000 mg/L	0.65533333
800 mg/L	0.56733333
600 mg/L	0.48166667
400 mg/L	0.3875000
200 mg/L	0.2980000

Figure 76. Average ABS readings, calibration curve, by concentration.

1000 mg/L	-0.6553
800 mg/L	-0.5673
600 mg/L	-0.4816
400 mg/L	-0.3875
200 mg/L	-0.2980

Figure 77. Data used for the determination of calibration curve.

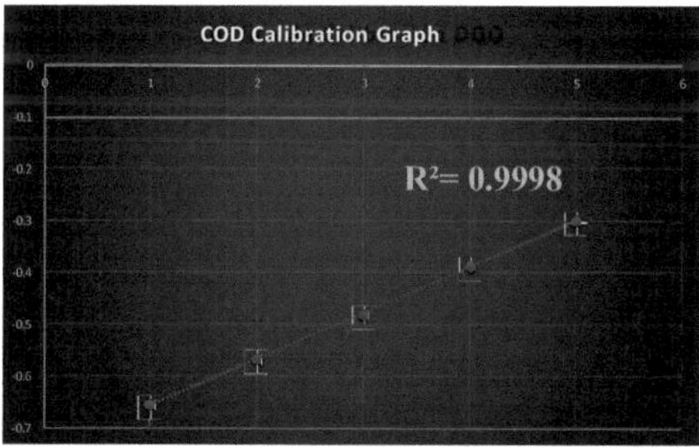

Figure 78. Calibration curve obtained, with deviation of 0.9998.

- By having the calibration curve, it was possible to carry out the determination tests on the water obtained in solar disinfection prototype, where the result was 0 mg of Oxygen /L of water, therefore the water did not contain organic matter.

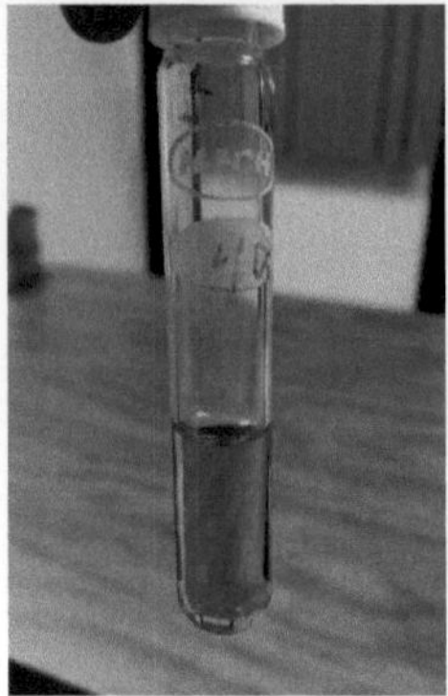

Figure 79. COD test result of the water obtained.

3.3 Measurement of dissolved solids and salts in natural, waste and treated waste water - test method.

In the validation process of the solar disinfection prototype for stagnant and rainwater, the measurement of solids in the treated water was carried out, following the NMX- AA-034- SCFI-2015 standard. This test was carried out with the aim of quantifying the amount of suspended and dissolved solids present in the water, which is essential to evaluate the effectiveness of the prototype in improving water quality and ensuring its suitability for human consumption.Sampling: This test was performed following the methodology of NMX- AA-034- SCFI-2015.

- The sample of wastewater treated by the prototype was collected.
- The sample was labelled and stored in clean containers for further analysis.

Figure 80. Water sample obtained.

Test specimen preparation: This test was performed following the methodology of NMX-AA-034-SCFI-2015.

- A 100 ml beaker, previously washed and dried, was used for the determination.

Figure 81. Measurement of solids per test tube.

Sample Filling: This test was performed following the methodology of NMX-AA-034- SCFI-2015.

- The water sample was filtered through a filter, where the filter retained the suspended solids present in the water, 100 ml of previously filtered water was placed.

Figure 82. Sample prepared for solids determination.

Sedimentation time: This test was performed following the methodology of NMX-AA- 034-SCFI-2015.

- The sample was left to stand for 24h.
- The sample was then capped to avoid any contamination or retention other solids.

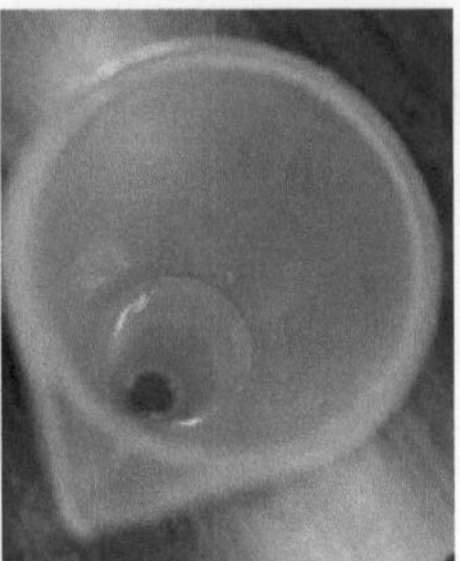

Figure 83. Water without sedimentation of solids is observed.

Measurement of Total Dissolved Solids (TDS): This test was performed following the methodology of NMX-AA-034-SCFI-2015.

- After it was left to stand, our sample was observed for solids.
- It was observed with the help of light, which did not show any solids. The result was excellent, the water is free of solids and no solid contamination.

Figure 84. Water completely free of solids is observed.

This test was conducted in order to assess solids in accordance with NMX-AA-034- SCFI-2015 and accurately evaluated the ability of the solar disinfection prototype to reduce solids in the treated water. This was crucial to ensure that the resulting water met the necessary quality standards for consumption, thus contributing to the improvement of the quality of life in communities with limited access to safe drinking water.

3.4 Water analysis pH measurement in natural, waste and treated waste water test method (NMX-AA-008-SCFI-2011).

In the validation of the solar disinfection prototype for stagnant and rainwater, the pH of the treated water was measured in accordance with the NMX-AA-008-SCFI-2011 standard. This test was carried out to determine the acidity or alkalinity of the water, a key parameter in the assessment of its potability. Maintaining an adequate pH is essential for ensure that the water is safe for human consumption and does not present corrosive or unhealthy conditions.
Sample Collection: This test was performed following the methodology of NMX-AA- 008-SCFI-2011.

- A water sample was collected from our prototype.
- The sample was stored in clean and labelled containers for proper identification.

Figure 85. Sample for alkalinity determination.

Calibration of the pH meter: This test was carried out following the methodology of NMX-AA- 008-SCFI-2011.

- Before starting the measurements, the pH meter was calibrated using buffer solutions of known pH (pH 4, pH 7 and pH 10) which ensured the accuracy of the readings.

Figure 86. Solutions for pH-meter calibration.

pH measurement: This test was performed following the methodology of NMX-AA-008-SCFI- 2011.

- The electrode of the pH meter was immersed in the water sample.
- The sample was gently shaken to ensure a homogeneous reading, and the value was allowed to stabilise.
- The pH value displayed on the pH meter for the sample was recorded and the reading was pH 6.6 which is a sample with the requested standards according to the regulations established by the Mexican nations.

Figure 87. pH meter used in the tests.

The importance of pH measurement according to NMX-AA-008-SCFI-2011 was crucial to assess the impact of the solar disinfection prototype on water quality. This test ensured that the treatment process did not significantly alter the pH of the water, keeping it within a safe range for human consumption.

3.5 Water analysis-measurement of electrical conductivity in natural, waste and treated waste water test method (NMX-AA- 093-SCFI-200).

In the validation process of the solar disinfection prototype for stagnant and rainwater, the electrical conductivity of the treated water was measured according to the NMX-AA-096-SCFI-2000 standard. This test was carried out in order to evaluate the concentration of dissolved ions in water, such as salts and minerals, which influence its ability to conduct electricity. Determining electrical conductivity is essential to ensure that treated water maintains adequate quality and is not affected by dissolved contaminants.

Sample Collection: This test was performed following the methodology of NMX-AA- 096-SCFI-2000.

• The sample was first labelled appropriately and stored in clean containers for analysis, the sample was extracted from the solar disinfection prototype.

Figure 88. Sample for determination electrical conductivity.

Calibration of the conductivity meter: This test was carried out following the methodology of NMX- AA-096-SCFI-2000.

• The measurements were carried out, the conductivity meter was calibrated using deionised water with a value of 0.00 mS, which guaranteed the precision of the instrument.

Figure 89. Calibration of conductivity meter.

Conductivity measurement: This test was carried out following the methodology of NMX-AA-096-SCFI-2000.

• The electrode of the conductivity meter was immersed in the water sample.

- The sample was gently agitated to obtain a stable and representative reading.
- The electrical conductivity value, expressed in microsiemens per centimetre (μS/cm), was recorded for each sample. Our sample obtained a value of 0.36 mS, so our sample is within the electrical conductivity parameters.

Figure 90. Prototype sample reading.

The measurement of electrical conductivity according to NMX-AA-096-SCFI-2000 was essential in which the quality of the water treated by the solar disinfection prototype was evaluated. This test made it possible to ensure that the disinfection process did not increase the concentration of dissolved ions in the water, maintaining its quality and potability within parameters suitable for human consumption.

CONCLUSIONS

The design and development of the solar disinfection prototype for stagnant and storm water represented a significant breakthrough in the search for sustainable water treatment solutions for communities with limited access to basic resources. Throughout the project, challenges inherent to the lack of infrastructure and financial resources in rural and peri-urban areas were addressed, leading to the development of an affordable and efficient system that uses solar energy as the main source of power. This approach not only aligned with sustainability principles, but also offered a practical solution to improve water quality in areas where access to safe drinking water is extremely limited. The verification tests performed, including Chemical Oxygen Demand (COD), which is free of organic matter, measurement of suspended solids, pH and electrical conductivity, provided highly satisfactory results. These tests showed that the prototype was able to significantly reduce the levels of contaminants in the treated water, meeting the quality standards required for human consumption. The system's ability to maintain a pH of 6.6, reduce the concentration of solids and ensure a low electrical conductivity (0.36 mS), confirmed its effectiveness in producing safe water suitable for various domestic and community applications.In addition, the evaluation of the ergonomics of the prototype was a key aspect to ensure its feasibility in the target communities. The final design was found to be not only easy to operate and maintain, but could also be adapted local conditions at a low implementation cost. This makes the prototype a replicable and scalable solution, with significant potential for adoption in other regions with similar characteristics. In summary, the project successfully met its objectives of offering an economical, efficient and sustainable solution for the treatment of stagnant and storm water in vulnerable communities. The results obtained not only validated the technical effectiveness of the prototype, but also demonstrated its positive impact on improving the quality of life of people living in areas with poor access to drinking water. This prototype represents an important step towards the resolution of one of the most critical problems in many regions of the world, providing a valuable tool to combat the scarcity of safe water and contribute to the well-being of communities most in need.

GLOSSARY

Solar disinfection: *The process of removing pathogenic micro-organisms from water by exposure to ultraviolet (UV) radiation from the sun.*

Stagnant water: *Bodies of water that do not flow, such as ponds, pools or cisterns, which can accumulate pollutants and pathogens.*

Stormwater: Rainwater *that can be collected and stored for various uses, but may contain pollutants depending on surfaces and collection methods.*

Prototype: An *initial model of a device or system used to test and validate its design and functionality prior to full-scale production.*

Chemical Oxygen Demand (COD): A *measure of the amount of oxygen required to oxidise the organic matter present in water. It is an indicator of organic pollution.*

Suspended Solids: *Solid particles present in water that do not dissolve and can affect water quality and treatment.*

pH: A *scale that measures the acidity or alkalinity of water. A pH of 7 is neutral, lower values indicate acidity and higher values indicate alkalinity.*

Electrical conductivity: The *ability of water to conduct electricity, which depends on the concentration of dissolved ions. It is an indicator of the amount of salts and other minerals in the water.*

Ultraviolet (UV) radiation: A *type of electromagnetic radiation with wavelengths shorter than visible light. UV-C radiation, in particular, is effective for water disinfection by damaging the DNA of microorganisms.*

Pathogenic micro-organisms: *Microscopic organisms, such as bacteria, viruses and protozoa, that can cause disease.*

Sustainability: The *ability of a system or process to be maintained over time without depleting resources or causing damage to the environment.*

Potability: *Water quality that makes it safe for human consumption, meeting established health and safety standards.*

Infrastructure: The *set of elements and services necessary for the functioning of a society, such as water supply, electricity and sanitation.*

Economic Resource: *Money, goods and services that are used to produce and distribute products and services.*

Pollutants: *Substances or elements that are present in water and may be harmful to human health and the environment.*

Verification: *The process of checking and validating that a system or device complies with established requirements and standards.*

Efficiency: The *ability of a system or process to achieve the desired effect with the use of available resources.*

REFERENCES

• 100Cía en Casa (2012, November 25). The power of Fresnel lenses. 100 Company at Home.https://100ciaencasa.blogspot.com/2012/11/el-poder-de-las-lentes- fresnel.html

• A20183390_21926. (2019, December 9). Isaac Newton: Reflection and refraction of light. Medium.https://medium.com/@a20183390_21926/isaac-newton- reflexi%C3%B3n-y-refracci%C3%B3n-de-la-luz-2fb8052fdd79

• Alejandro León. (2020, June 11). Preocupa agua estancada en Canal Nacional. Reforma. https://www.reforma.com/preocupa-agua-estancada-en-canal- nacional/ar2645461

• Renewable Alternative (2016, December 7). Evacuated tube solar water heater. Renewable Alternative Renewable. https://alternativarenovable.blogspot.com/2016/12/calentador-solar-de-agua-de- tubos-al.html.

• American Cancer Society (2020). Radiation therapy for cancer. Retrieved from https://www.cancer.org/cancer/cervical-cancer/treating/radiation.html

• ArchDaily (2013, November 29). Giant mirrors reflect the winter sun in the Norwegian city of Rjukan. ArchDaily. https://www.archdaily.mx/mx/02- 305625/giant-mirrors-reflect-the-winter-sun-in-the-norwegian-city-of-rjukan.

• Area Technology (n.d.). Irradiance and irradiation. Technology Area. https://www.areatecnologia.com/electricidad/irradiancia-irradiacion.html

• Avila, J. (2017). Analysis and evaluation of the cosine efficiency of a polar parabolic trough collector: Application in a subtropical region of Argentina. Advances in Science e Engineering, 8(1), 13-22. https://revistas.usfq.edu.ec/index.php/avances/article/view/2283/2904

• Beamex. (n.d.). Unitsof temperatura y their conversions. Beamex. https://blog.beamex.com/es/unidades-de-temperatura-y-sus-conversiones

• Canaltic. (n.d.). Energy solar thermal energy. Canaltic. https://canaltic.com/blog/html/exe/energias/energia_solar_trmica.html

• Cedric Olivier (n.d.) How does a rainwater harvesting system work? Comunidad Feliz. https://www.comunidadfeliz.mx/post/como-funciona-un-sistema- de-captacion-de-agua-pluvial?a_b_test_blog=C

• Chen, F. (2011). Solar energy: Fundamentals and applications. Springer.

• Science UNAM. (n.d.). From the wood boiler to the solar heater: A sustainable option. Ciencia UNAM. https://ciencia.unam.mx/leer/768/del-boiler-de-lena-al-calentador- solar-a-sustainable-option.

• CK-12 Foundation (n.d.). Solar energy and latitude. CK-12 Foundation. https://flexbooks.ck12.org/cbook/ck-12-conceptos-de-ciencias-de-la-tierra-grados-6-8-en-en-espanol/section/7.12/primary/lesson/la-energ%C3%ADa-solar-and-la-latitude/

• Climate Science. (n.d.). Effect greenhouse effect advanced. Climate Science. https://climatescience.org/es/advanced-greenhouse-effect

• National Water Commission (CONAGUA). (2021). Norma Oficial Mexicana NOM- 001-SEMARNAT-2021. Retrieved from https://www.gob.mx/conagua

• Concentration Solar. (n.d.). Disk parabolic. Concentration Solar. https://concentracionsolar.org.mx/concentracion-solar/disco-parabolico

• Diehl, J. F. (2002). Safety of irradiated foods. CRC Press.

• Digital Books Pro. (n.d.). [Title of chapter or book]. Digital Books Pro. https://reader.digitalbooks.pro/content/preview/books/39121/book/OEBPS/Text/cha pter1.html

• Docenteca. (n.d.). Heat vs temperature: Activities. Docenteca. https://www.docenteca.com/Publicaciones/455-calor-vs-temperatura-actividades.html#google_vignette

• Duffie, J. A., & Beckman, W. A. (2013). Solar engineering of thermal processes. John Wiley & Sons.

• Duffie, J. A., & Beckman, W. A. (2013). Solar Engineering of Thermal Processes (4th ed.) John Wiley & Sons.

• Duffie, J. A., & Beckman, W. A. (2013). Solar Engineering of Thermal Processes

(4th ed.) John Wiley & Sons.

• Eco Solar Esp (n.d.). Concentrating photovoltaic technology. Eco Solar Esp. https://www.ecosolaresp.com/la-tecnologia-de-concentracion-fotovoltaica/

• Ecoinventions (2018, January 15). New generation of concentrating solar power particle receivers. Ecoinventos. https://ecoinventos.com/nueva- generacion-de-receptores-de-particulas-de-energia-particulas-de-energia-solar-por-concentracion/.

• Ed Kashi, Christina Nunez (2024, April 25). Water pollution is growing global crisis. Here's what you need to know. National Geographic Spain. Retrieved from https://www.nationalgeographic.es/medio- ambiente/contaminacion-del-agua

• Elemetrics (2024). LP-PYRHE-16. Elemetrics. https://elemetrics.mx/producto/lp- pyrhe-16/

• Solar Energy (n.d.). Solar tracking systems. Solar Energy.

https://www.energiasolar.lat/sistemas-de-seguimiento-solar/

• Eurostar Solar (n.d.). Concentrating solar power systems. Eurostar Solar. https://www.eurostar-solar.com/sistemas-solares-de-concentraci%C3%B3n.html

• Facilitator_Ped (n.d.). Energy transfer by convection [Figure]. SlideShare. https://es.slideshare.net/Facilitador_Ped/los-materiales-y-el-calor-parte-2#8

• Energy Factor (n.d.). Irradiance and irradiance: Difference. Energy Factor. https://www.factorenergia.com/es/blog/autoconsumo-electrico/irradiacion-e- irradiance-difference/.

• Farkas, J. (2016). Irradiation for better foods. Trends in Food Science & Technology, 49, 1-2. https://doi.org/10.1016/j.tifs.2016.01.017

• Ferrer, F. J. (n.d.). Factors that condition terrestrial insolation. University of La Laguna. https://fjferrer.webs.ull.es/Apuntes3/Leccion02/2_factores_que_condicionan_la_ins olacin_terrestre.html

• Ferrer, F. J. (n.d.). Types of energy into which solar radiation is transformed [Figure]. University of La Laguna.

• Fordecyt-IER UNAM. (n.d.). Compound parabolic concentrator. Fordecyt-IER UNAM. http://www.fordecyt.ier.unam.mx/html/concentradorParab%C3%B3licoCompuesto _1.html

• García, L. P., & Hernández, M. T. (2017). Rainwater harvesting and its potential for rural areas. Water Resources Management, 31(5), 1329-1342. https://doi.org/10.1007/s11269-017-1609-3

• GMD Sol (n.d.). Solar thermal energy III: Solar tower. GMD Sol. https://gmdsol.com/energia-termosolar-iii-torre-solar/

• Gupta, N., & Verma, S. (2020). Applications of irradiation in food and agriculture. Journal of Food Science and Technology, 57(4), 1010-1018. https://doi.org/10.1007/s13197-019-04113-3

• Hach (2024). Pyranometer Kipp & Zonen CMP10. Hach. https://latam.hach.com/piranometro-kipp-zonen-cmp10/product?id=63882547647#

• Héctor Rodríguez. (2022, May 16). Oxygen levels in temperate lakes are at in declining. Magazine National Geographic Magazine Spain. Retrieved from https://www.nationalgeographic.com.es/naturaleza/niveles-oxigeno-lagos- templados-estan-declive_16972

• Hogarsense. (n.d.). Collector solar collector Hogarsense. https://www.hogarsense.es/energia-solar/captador-solar-termico

• Horta, P., Mendes, J. F., & Monteiro, L. P. (2018). Applications of solar thermal energy in industry. Energy Procedia, 153, 503-508. https://doi.org/10.1016/j.egypro.2018.10.051

• Hsaini, Y. (2020). Analysis of energy efficiency in residential buildings using dynamic simulations. [Final degree thesis, Universidad Politécnica de Madrid]. Repositorio Institucional de la Universidad Politécnica de Madrid. https://oa.upm.es/68142/1/TFG_YASIN_HSAINI_AMMI.pdf

• Ingelcia (n.d.). Benefits of heat pump for heating in industry. Ingelcia. https://www.ingelcia.com/articulo-general/beneficios-de-la-bomba-de- heat-for-heating-in-industry/.

• Iqbal, M. (1983). An Introduction to Solar Radiation. Academic Press. Kalogirou, S. A. (2014). Solar energy engineering: Processes and systems. Academic Press.

• JAPAC. (2018, July 16). Benefits of rural rainwater harvesting system. JAPAC. https://japac.gob.mx/2018/07/16/beneficios-del-sistema-de-captacion-pluvial-rural/

• Jonathan Castellón (2022, August 16.). Navojoa: Removal of stagnant water in Tetanchopo begins. Expreso. Retrieved from https://www.expreso.com.mx/noticias/sonora/navojoa-inician-retiro-de-agua- estancada-en-tetanchopo/158522

• Kalogirou, S. A. (2014). Solar Energy Engineering: Processes and Systems (2nd ed.). Academic Press.

• Keeui (2021, February 15). Power tower system. Keeui. https://keeui.com/2021/02/15/sistema-de-torre-de-energia/

• Khan, F. M. (2017). The physics of radiation therapy. Lippincott Williams & Wilkins.

• Kipp & Zonen (2024). Figure 4.16. Linke-Feussner Actinometer, manufactured by Kipp & Zonen. ResearchGate. https://www.researchgate.net/figure/Figura-416- Actinometer-Linke-Feussner-actinometer-manufactured-by-Kipp-Zonen_fig23_311375862

• Lara, Fernando & Velázquez, Nicolás & Sauceda, Daniel & Acuña, Alexis (2012). Methodology for the Sizing and Optimisation of a Linear Fresnel Concentrator. Informacion Tecnologica. 24. 10.4067/S0718-07642013000100013.

• Leutz, R., & Suzuki, A. (2001). Nonimaging Fresnel Lenses: Design and Performance of Solar Concentrators. Springer.

• Lovegrove, K., & Stein, W. (2012). Concentrating solar power technology: Principles, developments, and applications. Woodhead Publishing.

• Made-in-China (n.d.). Large size optical Fresnel solar lens, diameter 1100mm, solar concentrator solar energy Fresnel lens for cooking, Fresnel PMMA spot lens. Made-in-China.https://es.made-in-china.com/co_dgchinaoptics/product_Large-Size- Optical-Fresnel-Solar-Lens-Diameter-1100mm-Solar-Concentrator-Solar-Energy- Fresnel-Lens-for-Cooking-Fresnel-PMMA-Spot-Lens_uogrorsooy.html

• Madrimasd. (2021, November 26). Solar energy and its evolution: a perspective towards th the future. Madrimasd. https://www.madrimasd.org/blogs/energiasalternativas/2021/11/26/134995

• Mariana Roucau (n.d.). How is formed the rain forms how does acid rain form? Pinterest. https://es.pinterest.com/pin/435019645270457821/visual-search/?x=16&y=16&w=532&h=316&cropSource=6&surfaceType=flashlight

• Masters, G. M. (2004). Renewable and Efficient Electric Power Systems. John Wiley & Sons.

• Mekhilef, S., Saidur, R., & Safari, A. (2011). A review on solar energy use in industries. Renewable and Sustainable Energy Reviews, 15(4), 1777-1790. https://doi.org/10.1016/j.rser.2010.12.018.

• Mendes, F., Busscher, H. J., & van der Mei, H. C. (2018). Antimicrobial effects of electron beam and gamma irradiation. International Journal of Antimicrobial Agents, 52(3), 356-362. https://doi.org/10.1016/j.ijantimicag.2018.03.023

• Meteorology online (n.d.). IRIS2: The ambitious European satellite project. Meteorología en Red. https://www.meteorologiaenred.com/iris2-el-ambicioso- proyecto-europeo-de-satelites.html.

• Miller, A., & Smith, B. (2019). Non-ionizing radiation: Microwave and UV applications. Journal of Applied Physics, 125(14), 143301. https://doi.org/10.1063/1.5088324

• Molins, R. A. (2001). Food irradiation: Principles and applications. John Wiley & Sons.

• Monteith, J. L., & Unsworth, M. H. (2013). Principles of Environmental Physics: Plants, Animals, and the Atmosphere (4th ed.). Academic Press.

• World Health Organization [WHO] (2017). Guidelines for drinking-water quality: Incorporating the first supplement to the fourth edition (4th ed.). Geneva, Switzerland: WHO. https://www.who.int/water_sanitation_health/dwq/guidelines/es/

• Osterholm, M. T., & Norgan, A. P. (2004). The role of irradiation in food safety. New England ournal of Medicine, 350(18), 1898-1901. https://doi.org/10.1056/NEJMp048028

• Pihl, E., & Boulay, M. (2012). Concentrating Solar Power: Technologies, Costs and Opportunities in the U.S. Market. National Renewable Energy Laboratory.

• Pihl, E., & Boulay, M. (2012). Concentrating Solar Power: Technologies, Costs and Opportunities in the U.S. Market. National Renewable Energy Laboratory.

• Price, H., Lüpfert, E., Kearney, D., Zarza, E., Cohen, G., Gee, R., & Mahoney, R. (2002). Advances in parabolic trough solar power technology. Journal of Solar Energy Engineering, 124(2), 109-125. https://doi.org/10.1115/1.1467922

• Proain (n.d.). Importance of solar radiation in seedling production. Proain. https://proain.com/blogs/notas-tecnicas/importancia-de-la-radiacion-solar-en-la- seedling-production.

• Rabl, A. (1985). Active Solar Collectors and Their Applications. Oxford University Press.

• RESSSPI. (n.d.). Technology from technology RESSSPI. https://www.ressspi.com/calorsolar/tecnologia

• S., S. & Del Río, Jesus. (2009). Compound parabolic concentrator: An opto-geometric description. Revista mexicana de física E. 55. 141-153.

• Ministry of the Environment and Natural Resources (SEMARNAT). (1997). Norma Oficial Mexicana NOM-003-SEMARNAT-1997. Retrieved from https://www.gob.mx/semarnat

• Secretaría de Salud.(2002).Norma Oficial Mexicana NOM 230-SSA1-2002. Retrieved from https://www.gob.mx/salud

• Secretaría de Salud. (2021). Norma Oficial Mexicana NOM 127-SSA1-2021. Retrieved from https://www.gob.mx/salud

• Shutterstock (2024). Vector scientific illustration of heat flow isolated on white background. Shutterstock. https://www.shutterstock.com/es/image-vector/vector- scientific-illustration-heat-flow-isolated-2333500931

• Smith, J. A. (2020). Water quality and public health: The challenges of stagnant water. Journal of Environmental Studies, 45(2), 123-135.https://doi.org/10.1016/j.envres.2020.108987

• Smith, J., Doe, J., & Brown, P. (2015). Applications of radiation in research. Radiation Research, 183(3), 285-292. https://doi.org/10.1667/RR14062.1

• SolarAnywhere (2021). Data field definitions. SolarAnywhere. https://www.solaranywhere.com/es/support/data-fields/definitions/

• Stull, R. B. (1988). An Introduction to Boundary Layer Meteorology. Kluwer Academic Publishers.

• Suzel Tunes (2020, October 27). The advance of stagnant waters. Revista Pesquisa FAPESP. https://revistapesquisa.fapesp.br/es/el-avance-de-las-aguas-estancadas/

• Tecpa (n.d.). Types of solar thermal power plants. Tecpa. https://www.tecpa.es/tipos-decentrales-termosolares/.

• The Morning Star G2 (2012, March 16). Parabolic trough technology. The Morning Star G2. https://themorningstarg2.wordpress.com/2012/03/16/tecnologia- cylindrical-parabolic/

• The Morning Star G2 (n.d.). Spotlight. The Morning Star G2. https://themorningstarg2.wordpress.com/tag/concentracion-puntual/

• Thermal Engineering (n.d.) What is Fourier's thermal conduction law? Definition. Thermal Engineering. https://www.thermal-engineering.org/es/que-es- la-ley-de-conduccion-termica-de-fourier-definicion/

• Heat Transfer UNEFA Punto Fijo (n.d.). Energy transfer by solar radiation solar radiation [Figure]. https://transferenciadecalorunefapuntofijo.wordpress.com/wp-content/uploads/2015/04/medidor-radiacion-pce-spm1-squema.jpg

• Tricanal. (n.d.). Types of gutters for water gutters. Retrieved from https://tricanal.com/pluviales

• Tyagi, V. V., Kaushik, S. C., & Tyagi, S. K. (2012). Advancement in solar photovoltaic/thermal (PV/T) hybrid collector technology. Renewable and Sustainable Energy Reviews, 16(3), 1383-1398. https://doi.org/10.1016/j.rser.2011.12.013

• United Nations Scientifi Committee on the Effects of Atomic Radiation (UNSCEAR). (2008). Sources and effects of ionizing radiation. United Nations.

• Wagner, M. J., & Gilman, P. (2011). Technical Manual for the SAM Physical Trough Model. National Renewable Energy Laboratory.

• Wited (n.d.). Temperature and heat. Wited. https://www.wited.com/temperatura-y- heat/

• Zarza, E., Valenzuela, L., León, J., Hennecke, K., Eck, M., Weyers, H. D., & Eickhoff, M. (2004). Direct steam generation in parabolic troughs: Final results and conclusions of the DISS project. Energy, 29(5-6), 635-644. https://doi.org/10.1016/j.energy.2003.09.034

Printed by Books on Demand GmbH, Norderstedt / Germany